PRAISE FOR

extraterrestrial

"Part graceful memoir and part plea for keeping an open mind about the possibilities of what is out there in the universe—in particular, life. Loeb is one of the more imaginative and articulate scientists around . . . Half memoir, half soaring monologue."
—*New York Times*

"Well-written and compelling."
—*Forbes*

"It's good to be skeptical of audacious ideas. But it's also important to be open to audacious possibilities, and to think through their ramifications. Avi Loeb's sumptuously written book will provoke you to think about the possibility of intelligent life elsewhere in the universe in new and stimulating ways."
—Sean Carroll, *New York Times* best-selling author of *Something Deeply Hidden: Quantum Worlds and the Emergence of Spacetime*

"A tantalizing, probing inquiry into the possibilities of alien life."
—*Kirkus Reviews*

" 'There are more things in heaven and earth than are dreamed of in our philosophy,' Hamlet tells Horatio. In this passionately argued, visionary book, astrophysicist Avi Loeb urges us to abandon the arrogant fantasy that we are the only sentient life form in the universe. He proposes that 'Oumuamua, an anomalous interstellar object first sighted on October 19, 2017, was a piece of extraterrestrial technological equipment. The clues, as Loeb carefully reviews them, are fascinating, and still more are his arguments for what they might reveal to us about our own brilliant, blinkered, and quite possibly doomed civilization."
—Stephen Greenblatt, Pulitzer Prize–winning author of *The Swerve: How the World Became Modern*

"Ingenious . . . Loeb's thought-provoking work of popular science will entertain those who wonder if humans are alone in the universe."

—*Publishers Weekly*

"In this well-written and accessible book, a leading astronomer gives a provocative and thrilling account of the search for extraterrestrial intelligence, with emphasis on his own imaginative ideas. Above all, Professor Loeb asks us to think big and to expect the unexpected."

—Alan Lightman, *New York Times* best-selling author of *Einstein's Dreams* and *Searching for Stars on an Island in Maine*

"Loeb makes a persuasive scientific argument about 'Oumuamua's otherworldly origins."

—*New York Magazine*

"An exciting and eloquent case that we might have seen a sign of intelligent life near Earth—and that we should search further. As a world-renowned scientist and an innovative thinker, Avi Loeb opens your mind to some of the most important questions facing us as humans and convinces you that scientific curiosity is key to our future success."

—Anne Wojcicki, CEO and cofounder, 23andMe

"Part survey of thrilling new discoveries, part memoir of a restless intellect and part polemical airing of grievances, this curious volume from Harvard astrophysicist Avi Loeb challenges readers—and Loeb's contemporaries in the sciences—to take seriously the likelihood that we are not alone in the universe."

—*Shelf Awareness*

extraterrestrial

The First Sign of Intelligent Life Beyond Earth

AVI LOEB

HarperCollins*Publishers*
Boston New York

First Mariner Books edition 2022

For information, address HarperCollins Publishers, 195 Broadway, New York, NY 10007.

Mariner
An Imprint of HarperCollins Publishers, registered in the United States of America and/or other jurisdictions.

www.marinerbooks.com

Designed by Chrissy Kurpeski

Library of Congress Cataloging-in-Publication Data has been applied for.
ISBN 9780358645535 (trade paper)
ISBN 9780358274551 (e-book)
ISBN 9780358393788 (audiobook)

Printed in the United States of America
24 25 26 27 28 LBC 6 5 4 3 2

To my three muses, Ofrit, Klil, and Lotem,
and everyone else out there . . .

Contents

Introduction

WHEN YOU GET A CHANCE, STEP OUTSIDE AND ADmire the universe. This is best done at night, of course. But even when the only celestial object we can make out is the noontime Sun, the universe is always there, awaiting our attention. Just looking up, I find, helps change your perspective.

The view over our heads is most majestic at nighttime, but this is not a quality of the universe; rather, it is a quality of humankind. In the welter of daytime concerns, most of us spend a majority of our hours attentive to what is a few feet or yards in front of us; when we think of what is above us, most often it's because we're concerned about the weather. But at night, our terrestrial worries tend to ebb, and the grandeur of the moon, the stars, the Milky Way, and—for the fortunate among us—the trail of a passing comet or satellite become visible to backyard telescopes and even the naked eye.

What we see when we bother to look up has inspired humanity for as far back as recorded history. Indeed, it has recently been surmised that forty-thousand-year-old cave paintings throughout Europe show that our distant ancestors tracked the stars. From poets to philosophers, theologians to scientists, we have found in the universe provocations for awe, action, and the advancement of civilization. It was the nascent field of astronomy, after all, that was the impetus for the scientific revolution of Nicolaus Copernicus, Galileo Galilei, and Isaac Newton that removed the Earth from the center of the physical universe. These scientists were not the first to advocate for a more self-deprecating view of our world, but unlike the philosophers and theologians who preceded them, they relied on a method of evidence-backed hypotheses that ever since has been the touchstone of human civilization's advancement.

• • •

I have spent most of my professional career being rigorously curious about the universe. Directly or indirectly, everything beyond the Earth's atmosphere falls within the scope of my day job. At the time of this writing, I serve as chair of Harvard University's Department of Astronomy, founding director of Harvard's Black Hole Initiative, director of the Institute for Theory and Computation within the Harvard-Smithsonian Center for Astrophysics, chair of the Breakthrough Starshot Initiative, chair of the Board on Physics and Astronomy of the National Academies, a member of the advisory board for the digital platform Einstein: Visualize the Impossible from the Hebrew University of Jerusalem, and a member of the President's Council of Advisors on Science and Technology in Washington, DC. It is my good fortune to work alongside many exceptionally talented scholars and students as we consider some of the universe's most profound questions.

This book confronts one of these profound questions, arguably the most consequential: Are we alone? Over time, this question has been framed in different ways. Is life here on Earth the only life in the universe? Are humans the only sentient intelligence in the vastness of space and time? A better, more precise framing of the question would be this: Throughout the expanse of space and over the lifetime of the universe, are there now or have there ever been other sentient civilizations that, like ours, explored the stars and left evidence of their efforts?

I believe that in 2017, evidence passed through our solar system that supports the hypothesis that the answer to the last question is yes. In this book, I look at that evidence, test that hypothesis, and ask what consequences might follow if scientists gave it the same credence they give to conjectures about supersymmetry, extra dimensions, the nature of dark matter, and the possibility of a multiverse.

But this book also asks another question, in some ways a more difficult one. Are we, both scientists and laypeople, ready? Is human civilization ready to confront what follows our accepting the plausible conclusion, arrived at through evidence-backed hypotheses, that terrestrial life isn't unique and perhaps not even particularly impressive? I fear the answer is no, and that prevailing prejudice is a cause for concern.

• • •

As is true for many professions, fashionable trends and conservatism when confronting the unfamiliar are evident throughout the scientific community. Some of that conservatism stems from a laudable instinct. The scientific method encourages reasonable caution. We make a hypothesis, gather evidence, test that hypothesis against the available evidence, and then refine the hypothesis or gather more evidence. But fashions can discourage the consideration of certain hypotheses, and careerism can direct attention and resources toward some subjects and away from others.

Popular culture hasn't helped. Science fiction books and films frequently depict extraterrestrial intelligence in a way that most serious scientists find laughable. Aliens lay waste to Earth's cities, snatch human bodies, or, through torturously oblique means, endeavor to communicate with us. Whether they are malevolent or benevolent, aliens often possess superhuman wisdom and have mastered physics in ways that permit them to manipulate time and space so they can crisscross the universe — sometimes even a multiverse — in a blink. With this technology, they frequent solar systems, planets, and even neighborhood bars that teem with sentient life. Over the years, I have come to believe that the laws of physics cease to apply in only two places: singularities and Hollywood.

Personally, I do not enjoy science fiction when it violates the laws of physics; I like science and I like fiction but only when they are honest, without pretensions. Professionally, I worry that sensationalized depictions of aliens have led to a popular and scientific culture in which it is acceptable to laugh off many serious discussions of alien life even when the evidence clearly indicates that this is a topic worthy of discussion; indeed, one that we ought to be discussing now more than ever.

Are we the only intelligent life in the universe? Science fiction narratives have prepared us to expect that the answer is no and that it will arrive with a bang; scientific narratives tend to avoid the question entirely. The result is that humans are woefully ill prepared for an encounter with an extraterrestrial counterpart. After the credits roll and we leave the movie theater and look up at the night sky, the contrast is jarring. Above us we see mostly empty, seemingly lifeless space. But appearances can be deceiving, and for our own good, we cannot allow ourselves to be deceived any longer.

• • •

In "The Hollow Men," his meditation on post–World War I Europe, the poet T. S. Eliot reflects: "This is the way the world ends / Not with a bang but a whimper." In a few words, Eliot captures the devastation of that conflict, which was, at that point, the deadliest in human history. But perhaps because my earliest academic love was philosophy, I hear more than despair in Eliot's evocative lines. I also hear an ethical choice.

The world will end, of course, and most decidedly with a bang; our Sun, now about 4.6 billion years old, will in about 7 billion years turn into an expanding red giant and end all life on Earth. This is not up for debate, nor is it an ethical matter.

No, the ethical question that I hear in Eliot's "The Hollow Men" focuses not on Earth's extinction, which is a scientific certainty, but on the less than certain extinction of human civilization – perhaps, indeed, all terrestrial life.

Today, our planet is careening toward a catastrophe. Environmental degradation, climate change, pandemics, and the ever-present risk of nuclear war are only the most familiar of the threats we face. In myriad ways, we have set the stage for our own ending. It could come with a bang or a whimper or both – or neither. At the moment, all options are on the table.

Which path will we choose? This is the ethical question between the lines of Eliot's poem.

"Not with a bang but a whimper." What if this metaphor about endings holds true for certain beginnings? What if an answer to "Are we alone?" presented itself, but it was subtle, fleeting, ambiguous? What if we needed to employ our powers of observation and deduction to their fullest extent in order to discern it? And what if the answer to this question held the key to the other question I just posed, about whether and how terrestrial life and our collective civilization will end?

• • •

In the pages that follow, I consider the hypothesis that just such an answer was given to humanity on October 19, 2017. I take seriously not just the hypothesis but also the messages it contains for humanity, the lessons we might glean from it, and some of the consequences that could follow from our acting on or not acting on those lessons.

While pursuing answers to the questions of science, from the origins of life to the origins of everything, might appear to be among the most arrogant of human endeavors, the chase itself is humbling. Measured by all dimensions, each human life is infinitesimal; our individual accomplishments are visible only in the aggregate of many generations of effort. We all stand on the shoulders of our predecessors—and our own shoulders must support the endeavors of those who will follow. We forget that at our peril, and theirs.

There is humility, too, in appreciating that when we struggle to make sense of the universe, the fault is in our comprehension, not in the facts or the laws of nature. I was aware of this from an early age, a consequence of leaning toward becoming a philosopher in my youth; I learned it anew during my early training as a physicist and came to appreciate it more fully as a somewhat accidental astrophysicist. In my teens I was particularly struck by the existentialists and their attention to the individual confronting a seemingly absurd world, and as an astrophysicist I am particularly aware of my life—indeed, all life—measured against the vast scale of the universe. I have found that when viewed with humility, both philosophy and the universe inspire hope that we can do better. It requires proper scientific collaboration across all nations and a truly global perspective—but we can do better.

I also believe that sometimes humanity needs a nudge.

If evidence of extraterrestrial life appeared in our solar system, would we notice? If we are expecting the bang of gravity-defying ships on the horizon, do we risk missing the subtle sound of other arrivals? What if, for instance, that evidence was inert or defunct technology—the equivalent, perhaps, of a billion-year-old civilization's trash?

• • •

Here is a thought experiment that I put to the undergraduate students who attend my freshman seminar at Harvard. An alien spaceship has landed in Harvard Yard and the extraterrestrials make it clear that they are friendly. They visit, have their photographs taken on the steps of Widener Library, and touch the foot of the statue of John Harvard, as so many terrestrial tourists do. Then they turn to their hosts and invite them to climb aboard their spaceship for a one-way trip to the aliens' home planet. It's a little risky, they acknowledge, but what adventure isn't?

Would you accept their offer? Would you take that trip?

Almost all of my students answer in the affirmative. At this point, I change the thought experiment. The aliens remain congenial, but now they inform their human friends that rather than returning to their home planet, they are going to travel past the event horizon of a black hole. Again, it's a risky proposition, to be sure, but the aliens have enough confidence in their theoretical modeling of what awaits them that they're willing to go. What the aliens want to know is: Are you ready? Would you take that trip?

Almost all of my students answer no.

Both trips are one-way trips. Both entail unknowns and risks. So why the different answers?

The most commonly stated reason is that in the first instance, my students would still be able to use their phones to share their experiences with friends and family back home, for although it might take light-years for the signals to reach Earth, they would do so eventually. However, a trip past the event horizon of a black hole ensures that no selfie, no text, no information, whether wondrous or not, would ever get through. One trip would produce Facebook or Twitter likes; the other was guaranteed not to.

At this point I remind my students that, as Galileo Galilei argued after looking through his telescope, evidence doesn't care about approval. This applies to all evidence, whether it is learned on a distant planet or on the other side of a black hole's event horizon. The value of information doesn't reside in the number of thumbs-ups it gets but in what we do with it.

And then I put to them a question that many Harvard undergraduates feel they have the answer to: Are we — that is, human beings — the smartest kids on the block? Before they can reply, I add: Look skyward and realize that your answer will depend a great deal on how you respond to one of my favorite questions — are we alone?

Contemplating the sky and the universe beyond teaches us humility. Cosmic space and time have vast scales. There are more than a billion trillion sun-like stars in the observable volume of the universe, and even the luckiest among us live for merely 1 percent of a millionth of the lifetime of the Sun. But staying humble should not prevent us from trying to get to know our universe better. Rather, it should animate us to raise our ambitions, ask difficult questions that challenge our presumptions, and then set about rigorously pursuing evidence rather than likes.

• • •

Most of the evidence this book wrestles with was collected over eleven days, starting on October 19, 2017. That was the length of time we had to observe the first known interstellar visitor. Analysis of this data in combination with additional observations establishes our inferences about this peculiar object. Eleven days doesn't sound like much, and there isn't a scientist who doesn't wish we had managed to collect more evidence, but the data we have is substantial and from it we can infer many things, all of which I detail in the pages of this book. But one inference is agreed to by everyone who has studied the data: this visitor, when compared to every other object that astronomers have ever studied, was exotic. And the hypotheses offered up to account for all of the object's observed peculiarities are likewise exotic.

I submit that the simplest explanation for these peculiarities is that the object was created by an intelligent civilization not of this Earth.

This is a hypothesis, of course—but it is a thoroughly scientific one. The conclusions we can draw from it, however, are not solely scientific, nor are the actions we might take in light of those conclusions. That is because my simple hypothesis opens out to some of the most profound questions humankind has ever sought to answer, questions that have been viewed through the lens of religion, philosophy, and the scientific method. They touch on everything of any importance to human civilization and life, any life, in the universe.

In the spirit of transparency, know that some scientists find my hypothesis unfashionable, outside of mainstream science, even dangerously ill conceived. But the most egregious error we can make, I believe, is not to take this possibility seriously enough.

Let me explain.

1

Scout

LONG BEFORE WE KNEW OF ITS EXISTENCE, THE object was traveling toward us from the direction of Vega, a star just twenty-five light-years away. It intercepted the orbital plane, within which all of the planets in our solar system revolve around the Sun, on September 6, 2017. But the object's extreme hyperbolic trajectory guaranteed it would only visit, not stay.

On September 9, 2017, the visitor reached its perihelion, the point at which its trajectory took it closest to the Sun. Thereafter, it began to exit the solar system; its speed far away—relative to our star, it was moving at about 58,900 miles per hour—more than ensured its escape from the Sun's gravity. It passed through Venus's orbital distance from the Sun around September 29 and through Earth's around October 7, moving swiftly toward the constellation Pegasus and the blackness beyond.

As the object sped back to interstellar space, humanity remained unaware of its visit. Oblivious to its arrival, we hadn't given it a name. If anyone or anything else ever had, we were—and remain—ignorant of what that might be.

Only once it was past us did astronomers on Earth glimpse our departing guest. We assigned the object several official designations, finally landing on one: 1I/2017 U1. But our planet's scientific community and the public would come to know it simply as 'Oumuamua—a Hawaiian name reflecting the geographical location of the telescope used to discover the object.

• • •

The islands of Hawaii are jewels in the Pacific Ocean that attract tourists from around the world. But to astronomers, they hold an additional allure: they are home to some of the planet's most sophisticated telescopes, a testament to our most advanced technologies.

Among Hawaii's state-of-the-art telescopes are the ones that make up the Panoramic Survey Telescope and Rapid Response System (Pan-STARRS), a network of telescopes and high-definition cameras located at an observatory atop Haleakala, the dormant volcano that forms most of the island of Maui. One of the telescopes, Pan-STARRS1, has the highest-definition camera on the planet, and since it came online, the system overall has discovered most of the near-Earth comets and asteroids found in the solar system. But Pan-STARRS has another distinction—it gathered the data that initially tipped us off to 'Oumuamua's existence.

On October 19, astronomer Robert Weryk at the Haleakala Observatory discovered 'Oumuamua in the data collected by the Pan-STARRS telescope, images that showed the object as a point of light speeding across the sky, moving too quickly to be bound by the Sun's gravity. This clue quickly led the astronomy community to agree that Weryk had found the first interstellar object ever detected in our solar system. Yet by the time we had come up with

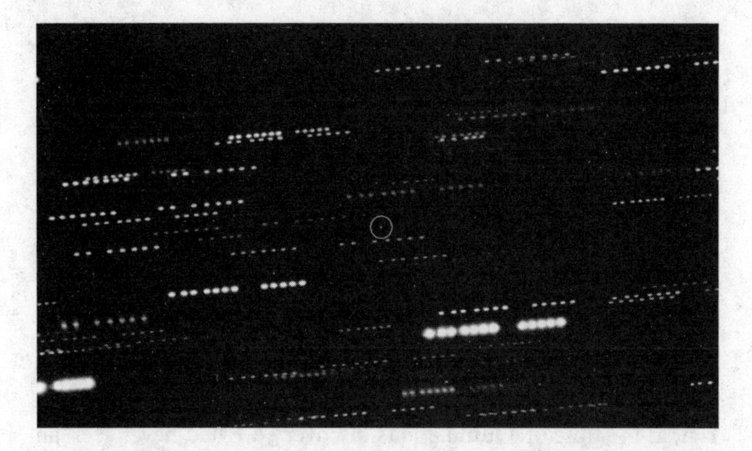

Combined telescope image of the first interstellar object, ʻOumuamua, circled, as an unresolved point source at the center. It is surrounded by the trails of faint stars, each smeared into a series of dots as the telescope snapshots tracked the moving ʻOumuamua. ESO/K. Meech et al.

a name for the object, it was over twenty million miles from Earth, or approximately eighty-five times as distant as the Moon, and rapidly moving away from us.

It came into our neighborhood a stranger, but it departed as something more. The object to which we had given a name had left us with a host of unanswered questions that would fully engage scientists' scrutiny as well as the world's imagination.

The Hawaiian word *ʻoumuamua* (pronounced "oh *moo* ah *moo* ah") is loosely translated as "scout." In its announcement of the object's official designation, the International Astronomical Union defined *ʻoumuamua* slightly differently, as "a messenger from afar arriving first." Either way, the name clearly implies that the object was the first of others to come.

• • •

Eventually, the media called ʻOumuamua "weird," "mysterious," and "strange." But compared to what? The answer, in brief, is that this scout was weird and mysterious and strange when compared to all other comets and asteroids previously discovered, ever.

In fact, scientists could not state with certainty whether this scout even *was* a comet or an asteroid.

It's not as if we didn't have a basis for comparison. Thousands of asteroids, dry rocks hurtling through space, are discovered every year, and the number of icy comets in our solar system is greater than our instruments can count.

Interstellar visitors are far rarer than asteroids or comets. In fact, at the time of ʻOumuamua's discovery, we had never seen an object that originated outside of our solar system pass through it.

This distinction was quickly lost. A second interstellar object was discovered shortly after ʻOumuamua was identified, and in the future, we are likely to discover many more, particularly with the Vera C. Rubin Observatory's upcoming Legacy Survey of Space and Time (LSST). And in a way, we had come to expect these visitors even before we could see them. Statistics suggest that while the population of interstellar objects crossing Earth's orbital plane is magnitudes smaller than the population of objects originating within the solar system, they are not themselves unusual. In short, the idea that our solar system sometimes plays host to rare interstellar objects is wondrous, but there is no mystery to it. And at first, the plainer facts of ʻOumuamua promised only wonder. Soon after ʻOumuamua's discovery was announced by the Institute for Astronomy at the University of Hawaii, on October 26, 2017, scientists around the world reviewed the most rudimentary data collected and agreed on most of the basic facts: ʻOumuamua's trajectory, its speed, and its approximate size (it was under one-quarter of a mile in diameter). None of these early details suggested that ʻOumuamua was unusual for any reason other than its origin outside our star system.

But before long, scientists sifting through the accumulating data began to point out ‘Oumuamua’s peculiarities—details that soon made us question the assumption that this object was a run-of-the-mill, albeit interstellar, comet or asteroid. Indeed, mere weeks after the object’s discovery, in mid-November 2017, the International Astronomical Union—the organization that names newly identified objects in space—changed its designation for ‘Oumuamua for the third and final time. Initially, the IAU had called it C/2017 U1; the *C* was for *comet*. Then it switched over to A/2017 U1; the *A* was for *asteroid*. Finally, the IAU declared it 1I/2017—the *I* stood for *interstellar*. By that point, that ‘Oumuamua had come from interstellar space was one of the few things upon which everyone agreed.

• • •

A scientist must go where the evidence leads, the old adage runs. There is humility in following the evidence, and it frees you from preconceptions that can cloud observations and insight. Much the same can be said for adulthood, a good definition of which might be “the point at which you have gathered enough experience that your models have a high success rate in forecasting reality.” Not, perhaps, how you would present it to your young children, but still, I find the definition has its virtues.

In practice, this simply means that we should allow ourselves to stumble. Let go of prejudices. Wield William of Occam’s razor and seek the simplest explanation. Be willing to abandon models that fail, which some inevitably do when they collide with our imperfect grasp of facts and the laws of nature.

Obviously, there is life in the universe; we are testament to that. And that means that humanity provides a vast, compelling, sometimes inspiring, and sometimes sobering data set to consider when we wonder about the actions and intentions of any other intelligent life that might exist—or might have existed—in the universe. As the only example of sentient life that we have studied in depth, humans are very likely to hold many clues to the behavior of any other sentient life, past, present, or future, in the universe.

As a physicist, I am struck by the ubiquity of the physical laws that govern our own existence on our particular little planet. When I look out into the cosmos, I am awed by the order, by the fact that the laws of nature that we find here on Earth seem to apply out to the very edges of the universe. And for a long time, since well before the arrival of 'Oumuamua, I have harbored a corollary thought: the ubiquity of these natural laws suggests that if there is intelligent life anywhere else, it will almost certainly include beings who recognize these ubiquitous laws and who are eager to go where the evidence leads, excited to theorize, gather data, test the theory, refine, and retest. And eventually, just as humankind has done, to explore.

Our civilization has sent five man-made objects into interstellar space: Voyager 1 and Voyager 2, Pioneer 10 and Pioneer 11, and New Horizons. This fact alone is suggestive of our unlimited potential to venture far out. So too is the behavior of our more distant ancestors. For millennia, humans have journeyed to the farthest reaches of our planet seeking different lives or better lives or just seeking, often with a shocking level of uncertainty about what they would find or whether they would return. Our species' certainty increased substantially over time—astronauts managed to travel to the moon and back in 1969—but the fragility of these undertakings remains. It wasn't the lunar module's walls, which were about as thick as a sheet of paper, that kept the astronauts safe; it was the science and engineering behind their construction.

And if other civilizations developed out there among the stars, wouldn't they have felt that same urge to explore, to venture past familiar horizons in search of the new? Judging by human behavior, that would not be surprising in the least. Indeed, perhaps these beings grew so comfortable with the limitless expanse of space that they traveled in it in much the same way that, here on Earth, we now traverse the planet. Our forebears used terms like *journeying* and *exploring*; today, we go on vacations.

In July of 2017 my wife, Ofrit, our two daughters, Klil and Lotem, and I visited an impressive collection of telescopes in Hawaii. As the chair of the Department of Astronomy at Harvard University, I had been invited to give a lecture on Hawaii's Big Island aimed at conveying the excitement of astronomy to the public, some of whom were protesting further construction of the next big telescope atop the dormant volcano Mauna Kea. I happily accepted and used the opportunity to visit some of the other Hawaiian islands, including Maui, that host state-of-the-art telescopes.

My subject was the habitability of the universe and the likelihood that in the coming decades, we would discover evidence of extraterrestrial life. And once we did, that discovery would force on humanity the appreciation that we're not that special. The local paper's headline covering my presentation nicely captured the idea: "Be Humble, Earthlings."

The lecture was given a little less than a month before 'Oumuamua—unbeknownst to Earthlings—passed the orbital plane of Mars, and I delivered it mere miles from Pan-STARRS1, one of the telescopes I visited on this trip and a technological marvel of instrumentation. Three months later, data gathered by Pan-STARRS would lead to the discovery of 'Oumuamua.

• • •

The first Pan-STARRS telescope, PS1, went online in 2008. Fifty years earlier, in 1958, another telescope had been built on the summit of Haleakala, but it was not used to study the stars; an animating fear at the time was Soviet satellites, and America wanted to be able to track them. Pan-STARRS, the Panoramic Survey Telescope and Rapid Response System, had a different objective: to detect comets and asteroids threatening to collide with Earth. As a consequence, since 2008 it has grown increasingly sophisticated. More telescopes have been added over the years, the most significant being Pan-STARRS2, which became fully operational in 2014. The array of telescopes collectively referred to as Pan-STARRS continue to map the skies above us, detecting comets, asteroids, exploding stars, and more.

In short, a bygone Cold War helped set in motion an observatory of such complexity and technological richness that, decades later, in the cold, clear atmosphere atop a dead volcano, a sophisticated instrument in the array was able to detect ‘Oumuamua, which passed overhead just a few years after this particular telescope opened for business.

It is easy to be impressed by the self-fulfilling quality of coincidences. But coincidences can be misleading. For much of human history, people have turned to mystical or religious explanations to make sense of occurrences that do not have clear causes. I like to think that even during our civilization's youth and early adolescence, humankind was gathering enough experience that its models had an increasing success rate in forecasting reality. Humanity, you might say, has slowly been entering adulthood over the course of recorded time.

In truth, most events in life stem from a confluence of multiple causes. This is true in casual examples (eating the soup in the bowl that sits before you) and in extraordinary cases (the origins of, well, everything). These can run from the very personal (say, the introduction that leads to the marriage that produces two daughters keen to vacation in Hawaii) to the global (say, the possibility—the very real possibility—that for eleven days in October of that year, our telescopes witnessed an object that originated outside the solar system).

• • •

My family and I returned from vacation to our century-old house outside of Boston, Massachusetts. It is in many respects vastly different from the farm in Israel where I was raised. But in the sense that it feeds my love of nature, my need to be in the midst of the things that grow and live among us, it is the same.

During an evening walk in the woods near my house, I witnessed a large tree falling in the forest that stretches out past our backyard. I first heard the cracks and then saw it give way and collapse. Its trunk, I saw, was hollow. Much of it had been dead for years, and on that date and at that time it could no longer hold up against the wind. It so happened I was there to see its demise – one part of a causal chain to which I was witness but over which I had no control.

But our actions can make a difference under more favorable circumstances. About a decade ago, when my family first moved to Lexington, I discovered a broken branch on a young tree in the yard. I was advised by a local gardener to cut off the nearly severed limb. On closer inspection, I saw that living fibers were still linking it to the rest of the tree. I chose to bind the branch together with insulation tape. Today the branch rises to the sky far above my head, but the insulation tape remains at eye level. That tree sits near the house, visible from our windows. I point it out to my daughters in order to remind them that humble acts can have extraordinary consequences.

Some of the most consequential decisions are made out of hopeful expectation of what might result. By the time I fixed the tree branch near my house, that was not only an article of faith for me but an oft-repeated experience.

2

The Farm

ONE OF MY EARLIEST MEMORIES IS OF ARRIVING A little late to school for my first day of first grade. When I walked into the classroom, the kids were running around and jumping on their chairs and even their desks. It was pandemonium.

My reaction was curiosity. I looked at my classmates and thought, *Should I join them? Does it make sense to behave like this? Why are they doing this? Why would I?* I stood by the door for a moment, trying to think my way through the questions.

The teacher came in a few seconds later. To say she was unhappy was an understatement. This was not how she wanted the new school year to begin. Attempting to assert her authority and calm down the students, she saw in me a chance to set things right. "Look at how well behaved Avi is," she said to the class. "Can't you all follow his example?"

But my placidity was not a sign of virtue. I hadn't decided that the right thing to do was to stand quietly and await the teacher's arrival; I just hadn't figured out whether it would make sense for me to join in the mayhem.

I wanted to tell the teacher this but did not, which I now think was unfortunate. The lesson my classmates might have learned from my behavior—a lesson I eventually learned myself and that I have since tried to teach my own students—wasn't about whether you should or shouldn't follow the crowd but rather that you should take time to figure things out before acting.

In deliberation, there is the humility of uncertainty. This, too, is an attitude toward life that I have worked to embrace, cultivate in my students at Harvard, and instill in my daughters. After all, it is what my parents sought to instill in me.

• • •

I grew up in Israel on our family's farm in Beit Hanan, a village about fifteen miles south of Tel Aviv. It is an agricultural community dating back to 1929, and shortly after its founding it boasted 178 inhabitants. By 2018, however, that number had increased to only 548. When I was a child, the village was defined by its orchards and greenhouses, which grew all kinds of fruits, vegetables, and flowers. It was also a moshav, a special type of village. Unlike a kibbutz, where land is farmed communally, a moshav consists of individual families who own their own farms.

Our farm was notable for its large field of pecan trees—my father was head of Israel's pecan industry—but we also grew oranges and grapefruit. When I was young, the pecan trees, which can grow to over one hundred feet, towered over me, but the citrus trees, with their distinctive, sharp odor when the fruit was ripe, rarely got above ten feet and were easier to climb.

Tending the groves and overseeing the necessary machinery was a full-time occupation for my father, David, who was a skilled problem solver. Indeed, I remember him most through objects: the tractors he maintained, the trees of our orchards he nursed, the appliances he mended throughout our home and farm. A particularly clear memory I have was his climbing atop the roof of our house in the summer of 1969 to ensure that the reception for our television would allow us to watch Apollo 11's lunar landing.

No matter how able my father was, the sheer extent of work meant that there remained plenty of daily chores for my two sisters and me. We raised chickens, and at a very young age I collected eggs every afternoon and spent many nights with a flashlight hunting down fluffy chicks that had escaped from their cages.

Israel in the 1960s and 1970s, the decades of my earliest years, was a precarious place. After World War II, Jewish refugees increased the population by about a third, and the number of people in the region went from two million to just over three million. Many came from Europe, and the echoes of the Holocaust were never absent. What's more, the Arab countries of the Middle East were resolutely hostile toward Israel, which was committed to holding its ground. One conflict followed another: the Sinai War of 1956 was followed by the Six-Day War of 1967, which was followed by the Yom Kippur War of 1973. Though only decades old at the time of my childhood, Israel was steeped in recent and ancient history, and Israelis then — as now — were aware that their nation's continued survival depended on deliberating over the consequences of their choices.

It is also a beautiful country, and Beit Hanan and my family's farm were splendid places to grow up in. This free atmosphere inspired my early writings, notes I collected and piled in the top drawer of my desk. Indeed, for much of my adulthood it was an animating faith for me that if my freethinking ways ever got me into trouble, I could always, and very happily, return to the farm of my childhood.

It is commonly thought that life is a collection of the places you visit. But this is an illusion. Life is a collection of events, and these are the results of choices, only some of which are ours to make.

There are, of course, continuities. The science I do is connected by a direct line to my childhood. It was an innocent time of wondering about the big questions in life, enjoying the beauty of nature, and, among the orchards and the close neighbors of Beit Hanan, not caring about my status or standing.

• • •

The chain of causation that brought me to Beit Hanan began, proximately, with the decision of my grandfather (and, in Hebrew, my namesake) Albert to flee Nazi Germany. More clear-eyed than many, he foresaw the likelihood of cataclysm, the fast-moving drift of events that, even before the outbreak of the Second World War, promised an ever-narrowing range of choices for Jews, an ever-growing risk of dire consequences if he did not select the right path.

Luckily for him, and for me, Albert made the right choice. He left Germany in 1936 and moved to Beit Hanan shortly after its founding. Although it was largely unsettled and, like the world, buffeted by the rising winds of war, the farming community was a comparatively safe haven. Soon after his arrival, he was joined by my grandmother Rosa and their two sons, one of whom was my father, then age eleven. When he transitioned from a German to a Jewish society, his name was changed from Georg to David.

My mother, Sara, came to Beit Hanan from afar as well. She was born and raised in Haskovo, near the Bulgarian capital of Sofia. The coincidence of geography that made her a Bulgarian and not a German saved her and her family during the war; although it allied itself with the Nazi regime, Bulgaria retained its sovereignty and thus some of its ability to resist Adolf Hitler's mounting demands for the country to deport its Jews to Germany. As the rumors of death camps circulated, the Bulgarian Orthodox Church protested deportations, and the Bulgarian king summoned the resolve to refuse Germany's requests. To be clear, he did so by declaring that Bulgaria needed its Jews for its own labor pool, but the consequence was that he managed to protect many of the nation's Jews. My mother was therefore able to enjoy a relatively normal childhood. She studied at a French monastery school and eventually entered college in Sofia. But in 1948, with postwar Europe a ruin and the Soviet Union expanding westward, she left school and emigrated with her parents to the new nation of Israel.

Beit Hanan's earliest founders were from Bulgaria, so the fact that Sara's family ended up there was not surprising. But the farm village was very different from the cosmopolitan city and university studies she had left behind. Her new home had its charms, however. Shortly after her arrival, Sara met my father. They fell in love, got married, and had three children – my two older sisters, Shashana (Shoshi) and Ariela (Reli), and, finally, me in 1962.

In those early years, my mother devoted herself to her family and the community. She was a locally renowned baker, and my wardrobe attested to her talent for knitting sweaters, but even in the relative isolation of Beit Hanan, she remained dedicated to a life of the mind. By this I mean not just a bookish interest in scholarship but a desire to apply her intellect to the world. And it was this, and her integrity, that caused her balanced judgments to be trusted by everyone who knew her, from the leadership of our village to visitors who came to our farm seeking her advice. I was a direct and daily beneficiary. She made clear how much she cared about my path in life, my choices and interests. Like a gardener watering and nurturing a plant, she was dedicated and meticulous in cultivating her children's curiosity.

She also followed her own. When I was a teenager, she went back to university and completed her undergraduate degree. She then went on to graduate school, earning a PhD in comparative literature. These were not undertakings that kept her at a remove from us; indeed, at her encouragement, I sat in on her undergraduate philosophy classes, and at her urging I worked my way through many of the books on her reading lists.

It was my mother who caused me to fall in love with philosophy, especially existentialism. I dreamed of thinking for a living. On weekends, I would grab a work of philosophy, most often something by the existentialists, including the novels they wrote and inspired, and then, with my chosen book, I'd drive our tractor to a quiet spot in the hills and read for hours.

• • •

It has occurred to me since these halcyon days on my family's farm that if humanity ever finds a habitable planet on which to establish an outpost of our civilization, the people who populate it will likely look and act much like the people of Beit Hanan. As human history shows, the immediate demands of settling outposts of a civilization recur.

Of necessity, they would focus on the growing of food and the collective effort of supporting one another, from eldest to youngest. They would each need to be resourceful and multitalented, capable of repairing and engineering machinery, cultivating crops, and educating the young. I believe, too, that they would accommodate the life of the mind, even in their remoteness. And, I suspect that when their children came of age, they would be met with the same expectation I was: mandatory service to society.

My plan to become a philosopher and address some of the fundamental questions that humanity had grappled with for eons was delayed owing to Israel's conscription of all citizens over the age of eighteen. Service was expected of everyone. Because I had shown promise in physics during high school, I was selected for Talpiot, a new program in which two dozen recruits per year worked in defense-related research combined with intense military training. My academic ambitions had to be set aside; the study of Jean-Paul Sartre and Albert Camus, the existential philosophers whom I'd read in my youth, did not fit the new role that had been assigned to me. Focusing on the study of physics was the closest I would get to an intellectually creative activity during my years of military service.

Though we wore the uniform of the Israeli Air Force, we were introduced to all branches of the Israeli Defense Forces. We underwent basic infantry training, took combat courses in artillery and engineering, and were taught how to drive tanks, carry machine guns on night-long treks, and parachute out of planes. Thankfully, I was athletically fit, so the physical challenges were demanding but bearable. And alongside these responsibilities, I avidly embraced my academic studies at the Hebrew University in Jerusalem.

Talpiot mandated that we study physics and mathematics, which sounded close enough to philosophy, and studying anything at the university seemed far more exciting than slogging through the muck with a rifle on my back. Given the opportunity, I did my best to justify the government's faith in me. It was also at this time that I started to realize that while philosophy asked the fundamental questions, it often couldn't resolve them. Science, I was learning, might put me in a better position to pursue answers.

• • •

After three years of study and military training I was supposed to start working on an industrial or military project that had immediate practical applications. But I sought a more creative path, one that posed greater intellectual and research challenges. I visited a facility that was not on the official list of research destinations, then drew up an outside-the-box research proposal. By that time I had established a track record of accomplishments, both in the classroom and in military training, and the Talpiot brass approved my idea—at first on a three-month trial basis and eventually for the remaining five years of my required service stint, from 1983 to 1988.

My work quickly evolved in new directions, some of which the military found quite intriguing. Through the thrill of scientific innovation, I developed the theory for a novel scheme (leading to a patent) to use an electric discharge in propelling projectiles to higher speeds than achievable with conventional chemical propellants. The project grew to employ an entire department of two dozen scientists and was the first international effort to receive funding from the United States' Strategic Defense Initiative (SDI), also known as "Star Wars," the ambitious missile-defense concept announced by President Ronald Reagan in 1983.

At the time, the Cold War, the decades-old contest between the United States and the Soviet Union, between democracy and Communism, between West and East, seemed a fixed feature of world affairs. Both sides had amassed vast arsenals of nuclear weapons, sufficient to destroy each other many times over. The Doomsday Clock, the brainchild of the members of the Bulletin of the Atomic Scientists and intended to warn humanity of the likelihood of a man-made catastrophe, was almost always set at seven minutes to midnight.

SDI was one part of that much larger contest. It envisioned using lasers and other advanced weapons to destroy incoming enemy ballistic missiles, and though it was dissolved in 1993, it had a major political impact on hastening the end of the Cold War and collapse of the Soviet Union.

This work also formed the backbone of my PhD dissertation, which I completed when I was twenty-four. The subject was plasma physics, which concerns the most common of the four fundamental states of matter; it's the stuff of stars, lightning, and certain television screens. (In case you're wondering, my dissertation was entitled "Particle Acceleration to High Energies and Amplification of Coherent Radiation by Electromagnetic Interactions in Plasmas" —a much less catchy title than this book's, to be sure.)

• • •

Even with my PhD in hand, I was unsure what my next choice should or would be. I wasn't married to a career in plasma physics. There was always the attraction of returning to Beit Hanan. And a big part of me wanted to change course dramatically and return to philosophy. However, a chain of choices, only some of which were mine, set me on a different path.

It started on a bus ride during my military service. The physicist Arie Zigler, seated next to me, happened to mention that the most prestigious place for postgraduate work was the Institute for Advanced Study (IAS) in Princeton, New Jersey. Later, during one of my visits to meet SDI officials in Washington, DC, and then at a plasma physics conference at the University of Texas at Austin, I crossed paths with the "pope of plasma physics," Marshall Rosenbluth. I knew that his past academic home was the institute, and I asked him for details. He quickly endorsed the idea that I go there for a short visit. Inspired, I immediately called Michelle Sage, the administrative officer at IAS, and asked if I could visit that coming week. She replied, "We do not allow just anyone to visit us. Please send me a copy of your CV and I will let you know if you can visit."

Undaunted, I mailed her a list of my eleven publications and called her again a few days later. This time she allowed me to schedule a visit at the end of my stay in the United States. When I arrived at her office early on the appointed morning, Michelle said, "There is only one faculty member here with available time, Freeman Dyson. Let me introduce you."

I was thrilled. I remembered Dyson's name from my textbooks on quantum electrodynamics. Soon after I sat down with Freeman in his office, he said, "Oh, you are from Israel. Do you know John Bahcall? He likes Israelis." He must have seen the curiosity on my face, for he elaborated: "His wife, Neta, is Israeli." I confessed that I had never heard of the man, let alone of his wife, Neta.

John Bahcall was an astrophysicist, I learned, and shortly thereafter I had lunch with him. It ended with him extending an invitation for me to return to Princeton for a monthlong visit. I learned that in the interim, he undertook an overseas reconnaissance, asking the most notable Israeli scientists, such as Yuval Ne'eman, what they thought of me. Whatever he was told, at the end of my second visit, John invited me to his office and offered me a prestigious five-year fellowship — but on the condition that I study astrophysics.

Of course I said yes.

• • •

When I was first encouraged to dedicate my professional career to astrophysics, I didn't even know what made the Sun shine. That Bahcall's area of expertise was the generation of weakly interacting particles called neutrinos in the Sun's hot interior made my ignorance about the topic all the more embarrassing. Up to that point, my focus had been on terrestrial plasmas and their more Earth-bound applications.

To be clear, Bahcall knew my area of past research. He extended the offer despite this fact. That he took the risk struck me as remarkable then and seems even more so now. (The state of academia has shifted since, and I doubt it would be possible to make a similar offer to a young scholar today.) I was and remain grateful. I accepted, determined to demonstrate that Bahcall's instincts — as well as the instincts of all the remarkable scientists who had helped me along this path — were justified.

Although I had to work to learn the basic vocabulary of the field so that I could begin to write original papers, the area was familiar. Plasma is a state that matter reaches at high temperatures when atoms are broken into a sea of positively charged ions (atoms that have lost some electrons) and negatively charged free electrons. Even though most of the ordinary matter in the present-day universe (including the interiors of stars) is in a plasma state, the field focuses on laboratory conditions, which are considerably different than those in space. Playing to my strengths, the first major research frontier that I pioneered as an astrophysicist centered on when and how the atomic matter in the universe was transformed into a plasma. So began my fascination with the early universe, the so-called cosmic dawn, or the conditions under which the very stars were formed.

After three years at IAS, I was encouraged to apply to junior faculty positions, including one at the Harvard astronomy department. I was their second choice. The department rarely offered tenure to junior faculty, so some candidates—including the person who was offered the job before me—thought twice about taking the position.

For my part, I gladly accepted. I recall thinking my decision through very clearly. I realized that if I wasn't offered tenure, I could always go back to my father's farm or take up my first academic love, philosophy.

I arrived at Harvard in 1993. Three years later, I received tenure.

• • •

I have since come to believe that John Bahcall not only had faith that I could handle the shift from plasma physics to astrophysics but also saw in me a kindred spirit—or perhaps even a younger version of himself. Bahcall had entered college intending to study philosophy but quickly concluded that physics and astronomy provided a more direct route to the most basic truths of the universe.

I came to a parallel realization not long after bidding farewell to John and the institute. When I took the junior faculty job at Harvard University in 1993, I decided it was too late to make a big career shift back into philosophy. More important, I had grown convinced that my "arranged marriage" to astrophysics had actually reunited me with my old love; it was just dressed up in different clothes.

Astronomy, I was coming to understand, addresses questions that were previously restricted to the realms of philosophy and religion. Among these questions are the biggest of the big—"How did the universe begin?" and "What is the origin of life?" I have also found that staring out into the vastness of space, contemplating the start and end of everything, provides a framework for answering, "What is a life worth living?"

Often the answer is staring us in the face. We need only summon the courage to admit it. During a visit to Tel Aviv in December of 1997, I had a blind date with Ofrit Liviatan. I liked her immediately, an appreciation that changed everything. Despite the geographical distance between us, we allowed our friendship to deepen. I had never met anyone like her and was convinced that I never would.

Long before I confronted the evidence presented by 'Oumuamua, I had learned that across all facets of life, taking the evidence presented to you and pursuing it with wonder, humility, and determination can change everything—if, that is, you are open to the possibilities contained in the data. Happily, by this point in my life, I was.

Ofrit and I were married two years later and, like me, she eventually found her place in the orbit of Harvard, serving as director of the university's freshman seminar program. In our old house near Boston, which was constructed just before Albert Einstein derived his theory of special relativity, Ofrit and I have raised our two daughters. That a causal chain runs from my grandfather's decision to leave Germany in 1936 to my parents' meeting in Beit Hanan to Ofrit and I raising Klil and Lotem in Lexington suggests to me that only a thin line separates philosophy, theology, and science. Watching children step slowly into adulthood reminds me that the most mundane acts of our existence suggest something miraculous that can be traced back to the Big Bang.

Over time, I have come to appreciate science slightly more than philosophy. Whereas philosophers spend a great deal of time inside their own heads, scientists are all about having a dialogue with the world. You ask nature a series of questions and listen carefully to the answers from experiments. When done frankly, it is a usefully humbling experience. The success of Albert Einstein's theory of relativity was not due to its formal elegance, which was developed over a series of publications from 1905 to 1915. It was not accepted until 1919, when Sir Arthur Eddington, secretary of the Royal Astronomical Society in England and an astronomer in his own right, confirmed the theory's prediction that the Sun's gravity would bend light. For scientists, what remains of a theory after its contact with data is what is deemed beautiful.

Although I am wrestling with the existential questions of my youth in a markedly different way than Jean-Paul Sartre or Albert Camus did, I believe the boy on the tractor in the hills of Beit Hanan would have been pleased with this outcome. He would have admired the sequence of opportunities and choices that started with a blind date and led to a family in Lexington.

But I understand, in a way that my younger self could not, another lesson of our family's story, one that I have kept at the front of my mind in recent years as I study interstellar visitors to our solar system.

Sometimes, by near accident, something exceptionally rare and special crosses your path. Life turns on your seeing clearly what's in front of you.

• • •

I believe that my life's unusual path prepared me for my encounter with 'Oumuamua. From a scientific standpoint, my experience taught me the value of freedom and diversity, specifically in the choice of research topics and the selection of collaborators, respectively.

The benefits of astronomers speaking with sociologists and anthropologists and political scientists and, of course, philosophers can be tremendous. Yet I have learned that in academia, interdisciplinary careers often share the fate of rare seashells swept onto the shore: if someone doesn't pick them up and preserve them, they erode over time until unrelenting ocean waves render them into indistinguishable grains of sand.

Throughout my own career, there have been many times when I could have been diverted to different, less fortuitous paths. My professional life has introduced me to many scholars who have qualifications similar to mine but who have not enjoyed opportunities similar to mine. An honest survey of faculty across academia brings to mind men and women whose contributions are defined by opportunities extended and opportunities taken away. The same can be said of nearly all walks of life.

Knowing that I have been the beneficiary of individuals who have extended such opportunities, I am deeply committed to helping young people fulfill their potential, even when that means challenging not just orthodox ideas but, sometimes, more pernicious orthodox practices. As part of this mission, I have worked hard to maintain—in my teaching and in my research—an approach to the world that some might consider childlike. If people think that, I won't take offense. In my experience, children follow their inner compasses more honestly and with fewer pretensions than many adults do. And the younger people are, the less likely they are to curb their thoughts to mirror the actions of others around them.

This approach to science has opened me up to some of the more ambitious—some might say *audacious*—possibilities inherent in the topics I study. For instance, the idea that 'Oumuamua, the interstellar object spotted tumbling across the sky in October 2017, was not a naturally occurring phenomenon.

3

Anomalies

SCIENCE IS LIKE A DETECTIVE STORY. FOR ASTROphysicists, this truism comes with a twist. No other field of scientific sleuthing confronts such a diversity of scales and concepts. Our chronological scope of inquiry starts before the Big Bang and stretches out to the end of time, even as we recognize that the very notions of time and space are relative. Our research descends to quarks and electrons, the smallest known particles; it reaches out to the edge of the universe; and it concerns — directly or indirectly — everything in between.

And so much of our detective work remains incomplete. We still don't understand the nature of the main constituents of the universe, and so out of ignorance we label them *dark matter* (which contributes five times more to the cosmic-mass budget than the ordinary matter we are made of) and *dark energy* (which dominates both dark and ordinary matter and that causes, at least at present, the peculiar cosmic acceleration). We also do not understand what triggered the cosmic expansion or what happens inside black holes—two areas of study in which I have been deeply involved since switching to astrophysics all those years ago.

There is so much we do not know that I often wonder whether another civilization, one that had the benefit of pursuing science for a billion years, would even consider us intelligent. The possibility that they might extend us that courtesy, I suspect, will not be determined by what we know but by how we know it—namely, our fealty to the scientific method. It will be in our open-minded pursuit of data that confirms or disproves hypotheses that humanity's claim to any universal intelligence will stand or fall.

Very often, what sets an astrophysicist's detective story in motion is the discovery of an anomaly in experimental or observational data, a piece of evidence that does not follow our expectations and that cannot be explained by what we know. In such situations, it is common practice to propose a variety of alternative explanations and then rule them out one by one based on new evidence until the correct interpretation is found. This was the case, for instance, with Fritz Zwicky's discovery of dark matter in the early 1930s; it was based on the observation that the motion of galaxies in clusters required more matter than was visible to our telescopes. His proposal was ignored until the 1970s, when additional data on the motion of stars in galaxies and the expansion rate of the universe provided conclusive evidence for it.

This winnowing process can divide, even fracture, whole fields of scholarship, pitting explanations and their advocates against one another until—sometimes—one side presents demonstrative proof.

This has been the case in the debate over 'Oumuamua, a debate that, for want of demonstrative proof, is ongoing. In fact, it is worth admitting up front that the likelihood of scientists ever obtaining demonstrative proof is very remote. Catching up to and photographing 'Oumuamua is impossible. The data we have is all we will ever have, leaving us the task of hypothesizing explanations that fully account for the evidence. This is, of course, a thoroughly scientific undertaking. No one gets to invent new evidence, no one gets to ignore evidence that is at odds with a hypothesis, and no one gets to—as in the old cartoon of a scientist working through a complex equation—insert "and then a miracle happens." Perhaps the most dangerous, most worrisome choice, however, would be declaring of 'Oumuamua, *Nothing to see here, time to move along, we've learned what we can and we'd best just go back to our old preoccupations.* Unfortunately, as of this writing, that seems to be what many scientists have decided to do.

The scientific debate over 'Oumuamua was relatively calm at the outset. I attribute this to the fact that early on, we were unaware of the object's most tantalizing anomalies. At first, this detective story seemed like an open-and-shut case: the likeliest explanation for 'Oumuamua—that it was an interstellar comet or asteroid—was also the simplest, most familiar one.

But as the fall of 2017 progressed, I, along with a significant portion of the international scientific community, found myself puzzling over the data. I—again, along with a significant portion of the international scientific community—couldn't make the evidence neatly fit the hypothesis that 'Oumuamua was an interstellar comet or asteroid. As all of us struggled to make the evidence fit that hypothesis, I began to formulate alternate hypotheses to explain 'Oumuamua's multiplying peculiarities.

• • •

Whatever else we conclude about 'Oumuamua, most astrophysicists would agree that it was, and remains, an anomaly unto itself.

For starters, prior to 'Oumuamua's discovery, no confirmed interstellar object had ever been observed in our own solar system. That alone made 'Oumuamua historic, and it was enough to draw many astronomers' attention, which led to the gathering of more data, which was interpreted and found to reveal further anomalies, which drew more astronomers' attention, and so on.

With the revelation of these anomalies, the real detective work began. The more we learned about ʻOumuamua, the clearer it became that this object was every bit as mysterious as the media reported.

As soon as the observatory in Hawaii announced its discovery, and even as ʻOumuamua was fleeing toward the outer solar system, astronomers around the world trained a variety of telescopes on it. The scientific community was, to put it mildly, curious. It was as if someone had come to your house for dinner and only when she was out the door and heading down the dark street did you become aware of all her strange qualities. We scientists had questions about our interstellar visitor and confronted a rapidly closing window of time to gather information, which we did by revisiting the data about our dinner guest that we had already collected and by observing her receding figure as she disappeared into the night.

One pressing question was: What did ʻOumuamua look like? We did not, and do not, have a crisp photograph of the object to rely on. But we do have data from all those telescopes that were dedicated for about eleven days to collecting whatever they could. And once we had our telescopes trained on ʻOumuamua, we looked for one bit of information in particular: how ʻOumuamua reflected sunlight.

Our Sun acts like a lamppost that illuminates not only all the planets orbiting it but every object that comes close enough to and is big enough to be seen from Earth. To understand this, you must first appreciate that in almost all scenarios, any two objects will rotate relative to each other when they pass. With that in mind, imagine a perfect sphere hurtling past the Sun as it makes its way through our solar system. The sunlight reflecting off its surface is unvarying, because the area of the tumbling sphere that faces the Sun is unvarying. Anything other than a sphere, however, will reflect the Sun's light by varying amounts as the object rotates. A football, for example, will reflect more light when one of its long

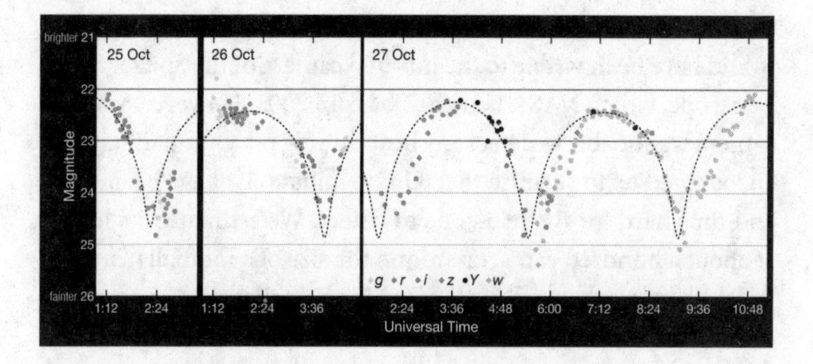

Variation in brightness of 'Oumuamua over time of day (in hours) as observed by different telescopes during three days in October 2017. The dots represent measurements through various filters in the visible and near-infrared bands of the color spectrum. The amount of reflected sunlight changed periodically by about a factor of ten (2.5 magnitudes) as 'Oumuamua rotated every eight hours. This implied that it had an extreme shape that was at least about five to ten times longer than it was wide when projected on the sky. The dashed white line shows the curve expected if 'Oumuamua were an ellipsoid with a 1:10 aspect ratio. Image by Mapping Specialists, Ltd. adapted from European Southern Observatory/K. Meech et al. (CC BY 4.0)

sides faces the Sun and less light when, as it tumbles, its narrow sides face the Sun.

For astrophysicists, an object's changing brightness provides invaluable clues to its shape. In the case of 'Oumuamua, the object's brightness varied tenfold every eight hours, which we deduced to be the amount of time that it took to complete one full rotation. This dramatic variability in its brightness told us that 'Oumuamua's shape was extreme, or at least five to ten times longer than it was wide.

To these dimensions, we added further evidence about 'Oumuamua's size. The object, we could say with certainty, was relatively small. Its trajectory near the Sun meant that 'Oumuamua should have had a very hot surface temperature, something that would have been visible to the infrared camera of the Spitzer Space Telescope, which NASA launched back in 2003. However, Spitzer's camera was unable to detect any heat coming off 'Oumuamua. This encouraged us to surmise that 'Oumuamua must have been small and thus hard for the telescope to detect. We estimated its length at about a hundred yards, or around the size of a football field, and its width at less than ten yards. Keep in mind that even a razor-thin object often appears to possess some width at a random orientation in the sky, so 'Oumuamua's actual width could well be smaller.

Let's assume that the larger of these dimensions is accurate and that the object measured a few hundred yards by a few tens of yards. This would make 'Oumuamua's geometry more extreme by at least a few times in aspect ratio—or its width to its height—than the most extreme asteroids or comets that we have ever seen.

Imagine setting down this book and taking a walk somewhere. You encounter other people. Perhaps they are strangers to you, and no doubt they all look different, but by their proportions, they are immediately recognizable as human. Among such passersby, 'Oumuamua would be a person whose waist appears to be smaller than his or her wrist. Seeing such a person would cause you to question either your vision or your understanding of people. This was essentially the dilemma that astronomers faced as they began to interpret the early data about 'Oumuamua.

• • •

As with any good detective story, the evidence that emerged about 'Oumuamua in the year after its discovery allowed us to abandon certain theories and winnow out hypotheses that did not fit the facts. Its brightness as it rotated gave us vital clues about what 'Oumuamua couldn't look like and what it might look like. In the latter category, the object's relatively small but extreme dimensions —with a length at least five to ten times greater than its width—

Artist's impression of 'Oumuamua as an oblong, cigar-shaped rock. This has become the dominant depiction of the interstellar object. ESO/M. Kornmesser

allowed only two possible shapes. Our interstellar visitor was either elongated, like a cigar, or flat, like a pancake.

Either way, 'Oumuamua was a rarity. If it was elongated, we had never seen any naturally occurring space object that size and that elongated; if it was flat, we had never seen any naturally occurring space object that size and that flat. Consider, for context, that all asteroids previously seen in the solar system had length-to-width ratios of, at most, three. 'Oumuamua's, as I have just noted, was somewhere between five and ten.

And there was more.

In addition to being small and oddly shaped, 'Oumuamua was strangely luminous. Despite its diminutive size, as it passed the Sun and reflected the Sun's light, 'Oumuamua proved to be relatively bright, at least ten times more reflective than typical solar system asteroids or comets. If, as seems possible, 'Oumuamua was a few times smaller than the upper limit of a few hundred yards that scientists presumed it to be, its reflectivity would approach unprecedented values — levels of brightness similar to a shiny metal.

• • •

When the discovery of ʻOumuamua was first reported, all of these peculiarities were arresting. Together, they presented a puzzle to astronomers. Together, they demanded a hypothesis that could explain why a naturally occurring object — and at this point, no one was arguing that ʻOumuamua was anything but — would have these statistically rare characteristics.

Perhaps, scientists reasoned, the object's strange features were caused by its exposure to cosmic radiation over the hundreds of thousands of years it had likely traveled in interstellar space before reaching our solar system. Ionizing radiation, in theory, could have significantly eroded an interstellar rock, although why such a process would have produced ʻOumuamua's shape isn't clear.

Or perhaps the reasons for its strangeness lay in ʻOumuamua's origin. Perhaps it had been violently expelled through a gravitational slingshot by a planet in a manner that explained some of its features. If a suitably sized object gets within a suitable distance of a planet, part of that planet could be pulled free and thrown, as by a slingshot, into interstellar space. Conversely, perhaps it was gently pulled free from a layer of icy objects orbiting the outer reaches of a solar system, something similar to our system's Oort cloud.

We could theorize a hypothesis starting from assumptions about ʻOumuamua's transit or from assumptions about its origins. If its peculiar shape and reflective properties had been the sum total of ʻOumuamua's distinctiveness, either theory might have been satisfactory. In that case, I would have remained curious but moved on.

But I could not restrain myself from joining in this detective story for one simple reason. It concerned 'Oumuamua's most arresting anomaly.

When 'Oumuamua sped part of the way around the Sun, its trajectory deviated from what was expected based on the Sun's gravitational force alone. There was no obvious explanation for why.

This, for me, was the most eyebrow-raising bit of data we accumulated over the roughly two weeks we were able to observe 'Oumuamua. This anomaly about 'Oumuamua, along with the other pieces of information that scientists assembled, would soon lead me to form a hypothesis about the object that put me at odds with most of the scientific establishment.

• • •

At one point during the frenzy that followed my articulation of my hypothesis about 'Oumuamua, I faced a roomful of reporters and a forest of outstretched microphones. I had just given three one-hour interviews. It was lunchtime; I was hungry. So rather than mount a detailed defense of my hypothesis about 'Oumuamua, I referred the journalists to one of my predecessors in the field of astronomy in the hope that doing so would encourage everyone to keep an open mind.

I reminded my audience of Galileo's seventeenth-century declaration that the evidence visible through his telescope suggested that the Earth orbited the Sun. It is one of the most familiar and oft-told stories in the annals of science: With the publication in 1610 of his treatise *Sidereus Nuncius* (which translates to *Starry Messenger*), Galileo described his observations of the planets via a new telescope and declared—based on this evidence—his agreement with the heliocentric theory of the solar system. Galileo's data implied that the Earth, along with all the other planets, revolved around the Sun. This ran directly counter to the teachings of the Catholic Church, which accused Galileo of heresy. Following a trial during which it is claimed his accusers refused to even look through his telescope, Galileo was found guilty of heresy. He spent the rest of his life, nearly a decade, under house arrest.

Galileo was forced to abandon his data and discovery and recant his statement that the Earth circled the Sun, but legend has it that afterward, Galileo whispered under his breath, "And yet it moves." The story is likely apocryphal, and even if it's true, its truth is beside the point—or at least it was for poor Galileo. Consensus had won out over evidence.

I didn't get into all of this at the press conference, of course; I just alluded to the story of the famous astronomer. But predictably, a reporter pounced: Are you saying you're Galileo? No. Not at all. What I wished to convey was that history has taught us to keep returning to the evidence about 'Oumuamua, testing our hypotheses against it, and, when others try to silence us, whispering to ourselves, "And yet it deviated."

• • •

To appreciate why 'Oumuamua's deviation was such an anomaly and why it led me to a hypothesis that has generated such intense controversy and pushback, we need to return to the basics. Let's recall one of the most fundamental of the physical laws that govern everything. It is Sir Isaac Newton's first law of motion: "Every object persists in its state of rest or uniform motion in a straight line unless it is compelled to change that state by forces impressed on it."

A billiard ball sits on a pool table, unmoving, even as fourteen other balls careen around it; it remains unmoving until another ball strikes it.

A solitary billiard ball sits on a pool table, unmoving, until a pool cue strikes it.

A billiard ball sits, stationary, on a pool table until someone lifts up one end of the table.

A billiard ball sits on a pool table, not moving, until a conical depression suddenly appears in the middle of the table.

In either of the last two cases, gravity takes hold and the ball starts to move. Once moving, it will do so in a line dictated by the force that acted on it and will continue traveling along that line until another force acts on it.

'Oumuamua entered our solar system on a trajectory that was roughly perpendicular to the orbital plane of Earth and the other planets. Just as the Sun exerts its gravitational force on those eight

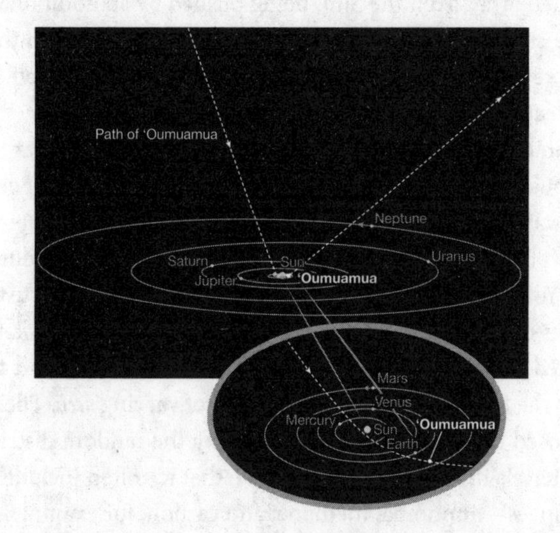

Trajectory of 'Oumuamua through the solar system, showing the location (inset) of the object and the planets on October 19, 2017, the date that 'Oumuamua was discovered by PanSTARRS. Unlike all asteroids and comets observed before, this object was not bound by the Sun's gravity. 'Oumuamua originated from interstellar space and returned there with a velocity boost as a result of its passage near the Sun. Image by Mapping Specialists, Ltd. adapted from European Southern Observatory/K. Meech et al. (CC BY 4.0)

planets and everything else orbiting it, the Sun exerted its gravitational force on ‘Oumuamua. On September 9, 2017, ‘Oumuamua sped around the Sun at almost 200,000 miles per hour, gaining momentum from the Sun’s gravitation, and then kicked in a different direction of motion. Thereafter it continued its journey through and beyond the solar system.

The universal laws of physics allow us to predict with certainty what a given object’s trajectory should be as it speeds around the Sun. But ‘Oumuamua didn’t behave as we expected.

In June 2018, researchers reported that ‘Oumuamua’s trajectory deviated slightly, but to a highly statistically significant extent, from a path shaped by the Sun’s gravity alone. This is because it accelerated away from the Sun, being pushed by an additional force that declined roughly as the square of the distance from the Sun. What repulsive force, which opposes the attractive gravitational force, can be exerted by the Sun?

Comets from the solar system show a deviation similar to ‘Oumuamua’s, but they are accompanied by a cometary tail of dust and water vapor from ice heated by sunlight.

If you have been lucky, you’ve seen a comet from your backyard. You’ve certainly seen photographs of comets or artists’ renderings of comets, their centers, or nuclei, fuzzily aglow and illuminated tails stretching out behind them. The glow and tail are due to the fact that comets are icy rocks of varying size. Their ice is composed mostly of water, but reflecting the random distribution of materials throughout the universe, that ice often includes other substances – ammonia, methane, and carbon, for example. Whatever the ice’s composition, it usually evaporates into gas and dust that scatter sunlight as the comet passes close to the Sun. This is what causes the comet’s coma, the enveloping atmosphere of evaporating ice and debris that gives a comet its glow and produces its distinctive tail.

If that tail reminds you of fuel coming out the back of a rocket, it should. The comet's evaporating ice acts like a jet that pushes the comet. Because of that rocket effect, an outgassing comet can deviate from a path shaped by the Sun's gravitation alone. Indeed, when astronomers observe such a comet, we can be precise. When we see an outgassing comet and measure the extent of its deviation, we can calculate how much of the comet's mass was used up in giving it this extra push.

If the extra push that propelled 'Oumuamua was from the rocket effect, as it is for comets, then our interstellar object should have lost a tenth of its mass in order for it to be propelled as much as it was. This is not a negligible amount of outgassing that could easily have been missed by our telescopes. But deep observations of the space around 'Oumuamua did not reveal any trace of water, carbon-based gases, or dust, ruling out the possibility that it was being pushed by cometary vapor or visible dust particles. Moreover, it did not change its spin rate as it should have if one-sided jets were pushing it sideways, as they often do in comets. Then, too, such a massive evaporation would have changed the tumbling period of 'Oumuamua, a phenomenon that is seen in solar system comets. No such change in the spin rate was recorded.

Ultimately, all of these mysteries can be traced back to one: 'Oumuamua's deviation from its expected path. All hypotheses as to what 'Oumuamua is have to account for that deviation, and that means explaining the force that acted on it while respecting the fact that if there was any cometary tail of gas and dust behind it, that tail was slight enough to go undetected by our equipment.

• • •

At the time of this writing, the scientific community has coalesced around the hypothesis that 'Oumuamua was a comet, albeit a peculiar one. A virtue of this hypothesis is its familiarity. We have observed many comets whose trajectories deviated from paths shaped by the Sun's gravity alone. We also know why that happens: in all cases, it is due to outgassing.

But as I have just explained, 'Oumuamua showed no outgassing. And yet it deviated.

We know that 'Oumuamua showed no outgassing visible to the infrared camera aboard the Spitzer Space Telescope. After its launch in 2003, Spitzer spent almost two decades circling about a hundred and fifty-five million miles above us while gathering an extraordinarily detailed body of information about our universe. And while its store of liquid helium, used to cool certain of its instruments to make them operable, was exhausted in 2009, its infrared array camera (IRAC) remained operational until January 2020, when it was finally taken offline.

The infrared camera on the Spitzer Space Telescope was ideal for surveying how much carbon dioxide comets produced. To an infrared camera, sufficient carbon dioxide is plainly visible. Because carbon is routinely part of comets' icy mix, and carbon dioxide is routinely the by-product of the evaporation of that mix when it is put under heat and stress, we frequently used Spitzer to observe comets' passing.

IRAC was trained on 'Oumuamua for thirty hours as our interstellar visitor sped past the Sun. Had there been even trace carbon dioxide in 'Oumuamua's outgassing, the camera should have been able to observe it. But IRAC saw nothing—not a trail of gas behind the object and certainly not the object itself. (Interestingly, the Spitzer Space Telescope did not detect any heat being emitted from 'Oumuamua either, implying that it must be shinier than a typical comet or asteroid; that is the only way it could have reflected as much sunlight as it did while still being small enough not to produce much heat.)

In a paper summing up their findings, the scientists who studied the IRAC data acknowledged that they "did not detect the object." However, they went on to state that "'Oumuamua's trajectory shows non-gravitational accelerations that are sensitive to size and mass and presumably caused by gas emissions."

Presumably. Having inserted that question mark in the middle of the sentence, the authors accurately concluded their article's abstract with the statement "Our results extend the mystery of 'Oumuamua's origins and evolution."

Other scientists using state-of-the-art equipment recorded results similar to the IRAC data. In 2019, astronomers reviewed images collected by the Solar and Heliospheric Observatory (SOHO) and the Solar Terrestrial Relations Observatory (STEREO) taken in early 2017 when 'Oumuamua was near perihelion (closest to the Sun). Built to observe the Sun, STEREO and SOHO were not intended to be comet finders (although after the latter identified its three thousandth comet, NASA declared it "the greatest comet-finder of all time"). Just like Spitzer, SOHO and STEREO did not detect anything in the area; to these instruments, 'Oumuamua was invisible. This can only mean that 'Oumuamua had a "water production rate" that was "smaller than any of the previously reported limits by at least an order of magnitude."

Invisible to Spitzer's IRAC, to SOHO, and to STEREO—and yet 'Oumuamua deviated.

• • •

To explain 'Oumuamua's trajectory and retain the assumption that it was a comet, scientists have strained to the breaking point their theories about its physical size and composition. For example, some scientists have hypothesized that 'Oumuamua's ice was entirely made of hydrogen, and this extreme composition explains why IRAC did not see it. (Outgassing containing carbon is visible to IRAC's infrared camera, but outgassing of pure hydrogen would not be.) In a detailed paper, my Korean collaborator Thiem Hoang and I calculated that a hydrogen iceberg traveling through interstellar space would evaporate long before it reached our solar system. As the lightest element in nature, hydrogen easily boils off an icy surface that is warmed up by interstellar radiation, gas and dust particles, and energetic cosmic rays. In fact, the periphery of the solar system is populated by numerous icy comets that are exposed to the same harsh environment (and the solar wind is unable to shield them since it is capped by the pressure of the interstellar medium much closer to the Sun). But a comet with ice composed of pure hydrogen—or, for that matter, pure anything—would be wildly exotic. We have never seen anything remotely like it before.

Or, rather, we know of nothing like it that is naturally occurring. To be sure, we have built such things; for instance, spacefaring rockets, for which pure hydrogen is the preferred fuel.

There is yet another difficulty with the outgassing-comet hypothesis, regardless of whether 'Oumuamua outgassed pure hydrogen or not. Its acceleration during deviation was smooth and steady. Comets are ungainly rocks; their rough and irregular surfaces retain unevenly distributed ice. As the Sun melts the ice and the outgassing produces propulsion, it does so across that rough and pitted surface. The result is what you would expect—a herky-jerky acceleration. But that is not what we saw 'Oumuamua do. In fact, it did the very opposite of that.

The odds of a naturally occurring comet composed of 100 percent hydrogen ice that outgasses from one location producing smooth acceleration? About the same as the odds of natural geological processes producing a space shuttle.

Moreover, to account for 'Oumuamua's degree of deviation, a statistically significant portion of its total mass would have had to be outgassed. The nongravitational push was substantial enough—about 0.1 percent of the Sun's gravitational acceleration—that cometary outgassing could be responsible for the deviation only if the process had expended at least 10 percent of 'Oumuamua's mass. That's a lot, and of course, this percentage represents a greater and greater amount of material the larger we hypothesize 'Oumuamua was; 10 percent of a thousand meters of stuff is more than 10 percent of a hundred meters of stuff.

Then, too, the more material we have to imagine 'Oumuamua invisibly outgassed, the less likely it becomes we would have failed to observe it. And the smaller we have to imagine 'Oumuamua was in order to explain why we did not see the material it outgassed, the odder its luminosity and width-to-length ratio becomes—and the shinier it would have had to be.

• • •

Outgassing isn't the only explanation for why an object would deviate from a path shaped by the Sun's gravity alone. Another explanation has to do with the object's disintegration.

If an object fractures, breaks up, and becomes smaller objects surrounded by dust and particles, the smaller objects follow a new trajectory. Thus, if 'Oumuamua began to break up around the time it reached the perihelion, that disintegration might have caused the object to deviate from the path dictated by the Sun's gravity.

The problem with using this explanation in the case of 'Oumuamua is that, just as with outgassing, our telescopes should have been able to register something—in this case, the relic fragments and dust from such a disintegration. It is unlikely that ice would have no carbon and even more unlikely that a disintegrating rock would contain no carbon. Further, one must wonder whether a collection of smaller objects would appear as a single body. 'Oumuamua, the evidence shows, continued to tumble every eight hours like a solid object with a persistent extreme shape.

The object's smooth acceleration also defies the hypothesis that 'Oumuamua fractured around perihelion, breaking up and losing enough of its mass to explain its deviated path. Our instruments observed no debris indicating such a fracturing and disintegration;

in fact, we saw evidence of the opposite: a smooth, steady acceleration. Had ‘Oumuamua started to break apart, the odds of it doing so while retaining smooth acceleration is, again, infinitesimal. Imagine a snowball thrown into the air that suddenly explodes into pieces but without any shift in the trajectory of the pieces.

For the disintegration hypothesis to hold, we are forced to make ever more exotic assumptions about ‘Oumuamua's composition to explain why we would not notice the vapor of the fragmented debris. Fragmentation should have increased what our instruments detected. After all, many small pieces of a disintegrating rock would increase the total surface area available, producing even more cometary gases and heat than the parent object alone.

And then there is the evidence that the extra force acting on ‘Oumuamua, the force that was causing it to deviate, declined in inverse proportion to the square of ‘Oumuamua's distance from the Sun. If the extra force were the result of outgassing, we would expect a faster deceleration of an object as it rapidly distances itself from the Sun. Evaporation of ice and water halts due to insufficient heating by sunlight, which ends the rocket effect. A rocket exhausts itself, and the extra force it was providing an object abruptly ceases; whatever path the object was on when that occurred is the path it thereafter follows. That is not what we saw ‘Oumuamua do. Again, the force acting on it declined in inverse proportion to the square of ‘Oumuamua's distance from the Sun.

What else could push ‘Oumuamua in this smooth power-law form? One possibility is the momentum delivered to ‘Oumuamua's surface by reflected sunlight. But for that to be effective, the surface-to-volume ratio needs to be unusually large. This follows from the fact that the solar push acts on the surface of the object, whereas the mass of an object (with some particular density of material) scales as its volume. Hence, the acceleration exhibited by the object increases in proportion to an increasing surface-to-volume ratio, which is maximized for an extremely thin geometry.

When I read reports that the extra force on ‘Oumuamua declined inversely with the distance from the Sun squared, I wondered what could be pushing it if not outgassing or disintegration. The only explanation that came to mind was the sunlight bouncing off its surface like wind off a thin sail.

• • •

Other scientists were busily crafting their own explanations. In pursuit of a theory that made sense of all the evidence, one scientist at the NASA Jet Propulsion Laboratory offered a new hypothesis that built on findings about the propensity for diminutive comets in nearly parabolic orbits to disintegrate right ahead of perihelion. Perhaps, he suggested, this was 'Oumuamua's fate. By the time it deviated from a trajectory determined by the gravity of the Sun, it had become a fluffy cloud of dust. Or, in his more precise language, 'Oumuamua became "a devolatilized aggregate of loosely-bound dust grains that may have exotic shape, peculiar rotational properties, and extremely high porosity, all acquired in the course of the disintegration event."

However loosely bound that cloud, this hypothesis still requires a devolatilized 'Oumuamua to be bound to some extent. After all, whatever remained had enough structural integrity that it was observed speeding away. *Devolatilization* means an object—say, a hunk of coal—is put under conditions, perhaps high heat, during which one element is removed. The example of devolatilization we are all familiar with is when a hunk of coal is heated to the degree that it becomes char.

This hypothesis holds that a comet not composed of carbon devolatilized into a highly porous bound exotic shape that was able to deviate to the statistically significant degree we observed 'Oumuamua deviate. And for that, it requires one more step. This structurally loosely bound dust cloud deviated without visible outgassing or debris from the "effects of solar radiation pressure."

A similar concept of an icy porous aggregate was advanced a

few months later by a researcher at the Space Telescope Science Institute. A decade earlier, this same scientist and I had collaborated to make the first prediction for the expected abundance of interstellar objects based on data for our solar system. (This prediction turned out to be orders of magnitude smaller than needed to account for 'Oumuamua, another implied anomaly.) Now my colleague wanted to explain the object's anomalous motion. In order for sunlight to produce the needed push, she calculated that the mean density of a porous 'Oumuamua had to be extraordinarily low, a hundred times more rarefied than air.

Just imagine an elongated cigar or pancake the size of a football field, sturdy enough to tumble every eight hours but so fluffy that it is a hundred times lighter than a cloud. This hypothesis strains plausibility, to put it mildly, not least because imagination is all we have to base it on, and we have never observed anything like it. Of course, the same is true of a naturally occurring cigar-shaped object or a naturally occurring pancaked-shaped object. We haven't seen such shaped objects, fluffy or not, at the extremes of 'Oumuamua.

Briefly ignore what the object is composed of and let's consider more carefully its shape. No one at a breakfast table would ever confuse a cigar for a pancake. They are dramatically different. So are we really left to choose between these two outlier shapes when we envision 'Oumuamua tumbling through space?

Yet another scientist, an astrophysicist at McMaster University, went back to the evidence to see if he could provide an answer. He evaluated all the brightness models the data allowed and concluded the likelihood of 'Oumuamua being cigar-shaped was small and the likelihood of 'Oumuamua being disk-shaped was about 91 percent. You should keep this percentage in mind when you see the umpteenth artist's rendering of 'Oumuamua as a cigar-shaped rock. You should also keep it in mind when reading any explanation for a naturally occurring oblong object, such as the low probability process of melting and tidal stretching along a rare trajectory that passes very close to a star, the value of which is mooted when it comes to ʻOumuamua, given this analysis.

Is there a simpler way to achieve the required surface-to-volume ratio for a pancake-shaped object? Yes, there is. You could build a thin, sturdy piece of equipment capable of deviating due to the effects of solar-radiation pressure to exactly such specifications.

4

StarChips

YEARS BEFORE 'OUMUAMUA'S DISCOVERY, I BEcame interested in the search for extraterrestrial civilizations and the possibility that Earth was not the only planet supporting life. It is an interest that stems from science and evidence rather than from science fiction. I love storytelling and I love science, but as I have confessed, I worry that narratives that violate the laws of physics and encourage a fascination with "improbables" get in the way of not only science but our own progress.

Anyway, who needs improbables when we have such a strong probable? The existence of intelligent life on Earth is more than sufficient justification to approach seriously the *scientific,* as opposed to fictional, search for life elsewhere in the universe.

I have felt this way since the start of my career in astrophysics. But this peculiar interest of mine became public only in 2007, when the cosmologist Matias Zaldarriaga and I proposed to eavesdrop on extraterrestrial radio signals.

It was a debut, of sorts—and it would prove to be transformative.

• • •

My unusual surveillance project with Matias grew out of my work on the early universe, the cosmic dawn that had attracted my attention back in 1993 when I moved from Princeton's Institute for Advanced Study to Harvard. A question that preoccupied me then was: When had the stars first "turned on"—that is, when was the moment the laws of nature declared, "Let there be light"? Contemplating the birth of stars would lead me, years later, to ponder how civilizations might eavesdrop on one another. But at the time, it was simply a question that I did not have the means to answer.

In brief, the effort to look far back in time to the earliest eons of the universe requires listening to the feeble radio emissions from primordial hydrogen, the most abundant element in the universe. This is best done with telescopes capable of searching for that early hydrogen's signature, an intrinsic wavelength of twenty-one centimeters that is stretched (shifted toward redder—longer—wavelengths, what is called "redshifted") to the scale of meters by the universe's expansion since the cosmic dawn.

By the mid-2000s, this theoretically possible field of experimental research was becoming a reality. Long-wavelength radio telescopes were at last under construction; one of them, the Murchison Widefield Array (MWA) in the desert of western Australia, was an international project involving scientists and institutions from Australia, New Zealand, Japan, China, India, Canada, and the United States.

As is true for so many of the world's observatories, the remote location of this kilometers-wide network of antennas was chosen for its lack of pollution—in this instance, not the absence of light pollution but the absence of radio waves broadcast by humans. Our televisions, cell phones, computers, and radios all emit radiation at the very frequencies to which the MWA telescope was tuned in its effort to pick up radio emissions from primordial hydrogen in the early universe—just another example of how technological advances can hinder rather than help astronomers.

All of this radio-wave pollution led me to a thought one day while I was eating my lunch alongside Matias and others. If our civilization emitted so much noise at that frequency, then perhaps other civilizations did too—extraterrestrial, alien civilizations that might exist out there among the very stars that Matias and I were studying.

It was an intuitive, spontaneous idea, one that initially elicited a laugh from Matias. But it became something more serious to both of us when I learned of the Foundational Questions Institute (FQXi) inaugural request for outside-the-box projects. With no baggage of past activity on related topics and resting on our reputations as mainstream scientists, I suggested to Matias that we turn this intriguing anecdote into an original research project. The fact that we were cosmologists not associated with the Search for Extraterrestrial Intelligence (SETI) Institute, which has always stood outside of the circle of more fashionable scientific organizations and has less advanced radio detectors and analysis, gave the project more credibility, and funding.

• • •

I have long been aware that within the discipline of astronomy, SETI faces hostility. And I have long found that hostility bizarre. Mainstream theoretical physicists now widely accept the study of extra-spatial dimensions beyond the three we are all familiar with—plainly put, height, width, and depth—and the fourth dimension, time. This is despite the fact that there is no evidence for any such extra dimensions. Similarly, a hypothetical multiverse—an infinite number of universes all existing simultaneously in which everything that could conceivably happen is happening—occupies many of our planet's most admired minds, again despite the fact that there is no evidence that such a thing is possible.

My complaint isn't with such endeavors; by all means, let theories multiply (and, perhaps, produce replicable experiments that provide supporting evidence). In fact, I take issue with the suspicion often visited on SETI. Compared to some flights of theoretical physics, the search elsewhere in the universe for something that is known to exist on Earth, the phenomenon of life, is a conservative line of inquiry. The Milky Way hosts tens of billions of Earth-size planets with surface temperatures similar to our own. Overall, about a quarter of our galaxy's two hundred billion stars are orbited by planets that are habitable in the way Earth is, with surface conditions that allow liquid water and the chemistry of life as we know it. Given so many worlds — fifty billion in our own galaxy! — with similar life-friendly conditions, it's very likely that intelligent organisms have evolved elsewhere.

And that's counting only habitable planets within the Milky Way. Adding all other galaxies in the observable volume of the universe increases the number of habitable planets to a zetta, or 10^{21} — a figure greater than the number of grains of sand on all of the beaches on Earth.

Some of the resistance to the search for extraterrestrial intelligence boils down to conservatism, which many scientists adopt in order to minimize the number of mistakes they make during their careers. This is the path of least resistance, and it works; scientists who preserve their images in this way receive more honors, more awards, and more funding. Sadly, this also increases the force of their echo effect, for the funding establishes ever bigger research groups that parrot the same ideas. This can snowball; echo chambers amplify conservatism of thought, wringing the native curiosity out of young researchers, most of whom feel they must fall in line to secure a job. Unchecked, this trend could turn scientific consensus into a self-fulfilling prophecy.

By limiting interpretations or placing blinders on our telescopes, we risk missing discoveries. Recall the clerics who refused to look through Galileo's telescope. The scientific community's prejudice or closed-mindedness—however you want to describe it—is particularly pervasive and powerful when it comes to the search for alien life, especially intelligent life. Many researchers refuse to even consider the possibility that a bizarre object or phenomenon might be evidence of an advanced civilization.

Some of these scientists claim that they simply will not dignify such speculation with their attention. But as I've noted, other forms of speculation are enshrined in the scientific mainstream—the existence of multiple universes, for example, and the extra dimensions predicted by string theory, and this despite the fact that there is no observational evidence for either of these ideas and perhaps never will be.

I will return to the subject of SETI and the academic community's resistance to it later in this book, since it is a topic that grows even more consequential once you understand the full scope of its implications. For now, suffice to say that in comparison to lots of mainstream scientific ideas, the search for alien life—even the intelligent variety—is not such a speculative endeavor. After all, a technological civilization emerged here on Earth, and we know that there are a lot of other planets like ours out there.

• • •

When Matias and I pursued our questions about eavesdropping on alien civilizations, it was less because we thought we would promptly hear the communications of such civilizations and more because we believed it would help direct attention and effort toward another question: Are we alone?

In the years that followed my endeavor with Matias, I was drawn to more and more SETI-related issues. What would be an evidence-backed approach to answering its guiding question? And because it joined a list of topics that aroused my curiosity—the nature of black holes, the beginnings of the universe, the possibility of near-light-speed travel—I found myself keeping company with any scholar whose interests overlapped my own, including some scientists who were exclusively identified with the search for alien intelligence.

Subsequently, Ed Turner, an astrophysicist at Princeton University, and I were the first to consider how one might go about seeking evidence of artificially produced light. We had the idea that we might try to find the glimmer of, say, a spacecraft or a city at great distances with our modern telescopes. Encouraged by Freeman Dyson, we then turned the question on its head and started to wonder whether you could see a city of the size and brightness of Tokyo on Pluto, which at that time was the most distant planet in our solar system. (It has since been reclassified as a dwarf planet.) Our proposition was more theoretical than practical; we never seriously considered focusing our telescopes on the icy ball that is Pluto in search of a city. Rather, our thought exercise was designed to figure out what we (or any civilization, for that matter) might do to seek out a city's telltale light signature among the twinkling stars.

It turns out that if you used an instrument of the technological sophistication of the Hubble Space Telescope and sought the signature of artificial light for long enough, you could indeed see Tokyo from the edge of the solar system. And you could tell that the light was intrinsic and not reflected sunlight based on how the source dimmed with increasing distance from the Sun.

By 2014, my reputation for taking seriously the question of whether or not we are alone in the universe had grown to the point that a writer from *Sports Illustrated* contacted me to address a fanciful notion raised by the president of FIFA: the possibility of an interplanetary World Cup. However tongue-in-cheek the original comment was, the magazine wanted someone to weigh in on the viability of such an idea. I gamely walked the writer through the various barriers, from the technology necessary to transport teams to the playing fields to the need to agree on atmospheric conditions of play, after pointing out the obvious: first, we needed to find intelligent life to compete with.

We were closer to that goal than I realized, for it was also around this time, and with far more serious purpose, that Yuri Milner sought me out.

• • •

A billionaire entrepreneur from Silicon Valley, Yuri Milner radiates an intensity of purpose. He was born in the Soviet Union, studied theoretical physics at Moscow State University, received his MBA from the Wharton School of the University of Pennsylvania, and became a stunningly successful investor. Companies he has helped to support include Facebook, Twitter, WhatsApp, Airbnb, and Alibaba.

In May of 2015, Yuri and Pete Worden, the former director of NASA's Ames Research Center, came by my office at the Harvard-Smithsonian Center for Astrophysics to encourage me to participate in a new program they were launching, a project they would come to call the Starshot Initiative. They wanted to support a team that would engineer and launch spacecraft capable of reaching the star system closest to ours: Alpha Centauri, a group of three stars orbiting one another some 4.27 light-years from Earth.

That Yuri would advance such an undertaking was not surprising. In 2012, he and his wife, Julia, had established the Breakthrough Prize. Every year, it awarded prize money to international scholars working in three fields: fundamental physics, life sciences, and mathematics; each prize was worth three million dollars. Within a year, Yuri and Julia were joined by the likes of Mark Zuckerberg, cofounder of Facebook; Sergey Brin, cofounder of Google; and Anne Wojcicki, cofounder of 23andMe, in supporting these awards.

By 2015, Yuri was thinking of other, more direct and ambitious ways of advancing scientific projects that excited him, so he launched the Breakthrough Initiatives. The focus was unambiguous. The project was to seek answers to two of the most fundamental questions confronting humanity: Are we alone? And can we, by thinking and acting together, make the great leap to the stars?

Yuri had been fascinated by these questions ever since, at a young age, he read the book *Universe, Life, Intelligence,* published in 1962 by the Soviet astronomer Iosif Shklovskii. (Shklovskii later released in an English edition, *Intelligent Life in the Universe,* coauthored with the American astronomer Carl Sagan.) It may have helped that Yuri's parents named him after the famed Soviet cos-

Artist's illustration of Proxima b, the nearest habitable planet outside of our solar system. This planet, discovered in August 2016, has a mass of roughly one to two Earth masses and revolves around Proxima Centauri, a dwarf star with 12 percent the mass of the Sun, at a distance of 4.24 light-years from Earth. Proxima b has a surface temperature comparable to that of the Earth, but because of its proximity to its faint host star, it is believed to be tidally locked, with permanent day and night sides. ESO

monaut Yuri Gagarin, who in 1961—the year of the younger Yuri's birth—became the first human to be launched into space.

In truth, I was prepared to offer Yuri my help even before he had fully framed his request. His bold and sincere interest in exploring whether there was life beyond Earth resonated completely with my perspective. Still, his expectations were daunting. Yuri explained that he wanted me to lead a project to send a probe to the three-star system Alpha Centauri to determine if there was life there. The catch was that it had to be done within Yuri's lifetime. I asked for six months to come up with the appropriate technological concept.

Working with my students and postdocs, I critically examined the options for meeting the goal of the Starshot Initiative. An attractive target within the Alpha Centauri system was Proxima Centauri, the star nearest Earth. To our joy, and just a few months after the announcement of Starshot, it was discovered that this dwarf star hosted a planet, Proxima b, in its habitable zone.

A chemical-propulsion rocket, which is what has sent all of our Earth-launched ships into space, would take approximately one hundred thousand years to reach Proxima b. Yuri was fifty-six, so given his stipulated timeframe—within his lifetime—a propulsion rocket was a nonstarter.

To get to Proxima b within several decades, we needed a spacecraft capable of traveling at a fifth of the speed of light. Even if we used nuclear fuel, which has the highest energy density of all fuels (other than antimatter, which is unavailable), it would be impossible for a propulsion rocket to reach such speeds. And Newton's second law of motion—which states that the acceleration of an object depends on its mass and the force acting upon it—dictated that our spacecraft would have to weigh as little as possible too.

To get an object to accelerate to the desired speed would take a tremendous amount of energy; the lighter the object, the less energy would be required. Correspondingly, our spacecraft's payload would have to be no more than a few grams. That led to another

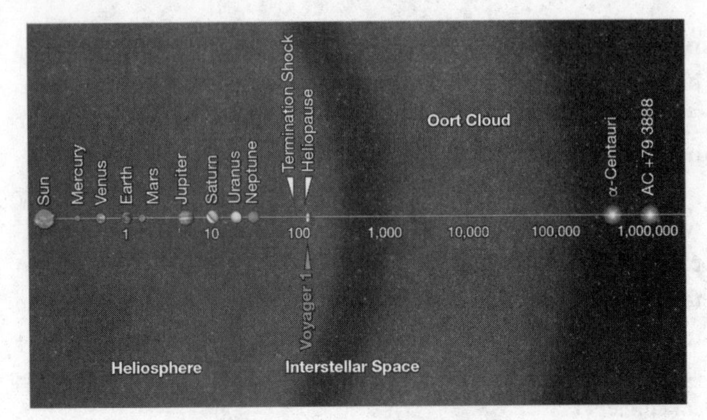

The journey to the nearest star system, Alpha Centauri, located about four light-years away, would take tens of thousands of years with a conventional chemical-propulsion rocket (if it had started when the first humans left Africa, it would be completing the trip now). The edge of the solar system is marked by the Oort cloud, stretching halfway to Alpha Centauri. Distances are labeled in units of the Earth-Sun separation (1 astronomical unit). In 2012, Voyager 1 crossed the heliopause, which is where the solar wind collides with interstellar gas. Image by Mapping Specialists, Ltd. adapted from NASA/JPL-Caltech

challenge. Not only would our spacecraft need to cover the vast distance in much less than one hundred thousand years, but once it reached Proxima Centauri, it had to be able to take pictures and send them back to Earth in a manner we could detect. It had to be light, small, and inexpensive to manufacture. This dictated that the camera and transmitter would be similar to what is in today's mobile phones. By our calculations, that technology, with some modifications, would suffice.

We discarded ideas and refined the ones left, and eventually we converged on a plan to launch a lightweight spacecraft attached to a reflecting sail—a mirror, essentially. The idea of a solar sail—a manufactured object that would be propelled by the pressure sunlight exerted on it—was centuries old. As early as 1610, Johannes

Kepler wrote to Galileo of "ships or sails adapted to the heavenly breezes." The feasibility of building such an object, however, did not become even remotely possible until the 1970s. For one thing, light transforms into heat when absorbed, as any dog or cat who finds a sun patch to nap in knows. So our mirror could not be just any mirror; it had to absorb less than one one-hundred-thousandth of the light striking it so that it would not burn up. And then we would need to hit that lightsail with an extremely powerful, extremely accurate laser.

This, too, was not an entirely original thought. The concept of a laser-propelled sail was invented the year I was born (1962) by the visionary Robert Forward and subsequently developed by other scientists, such as Phil Lubin, to include miniaturized electronics and modern optical designs. But never before had it been so close to becoming a reality.

We calculated that a 100-gigawatt laser beam capable of targeting a sail roughly the size of a person for a few minutes would propel the sail, along with its attached camera and communication device, to one-fifth of the speed of light by the time the spacecraft was five times as far away as the moon. This would be the spacecraft's open runway, so to speak. Across this distance, the laser beam could launch the spacecraft with enough speed to allow it to reach the nearest star within our lifetimes.

Everything we proposed was within existing technological bounds. Difficult? Yes. Expensive? Somewhat; it was on the order of one of the biggest science projects, like the Large Hadron Collider at CERN or the James Webb Space Telescope, but cheaper than the Apollo moonshot. (Many people who heard about the Starshot Initiative confessed that they hadn't been as excited about space exploration since the Apollo mission fifty years earlier.) But it would also be efficient. Once built, the launch system could be used to send lots of such craft—which are perhaps better thought of as probes, though we fell into the habit of referring to them as StarChips.

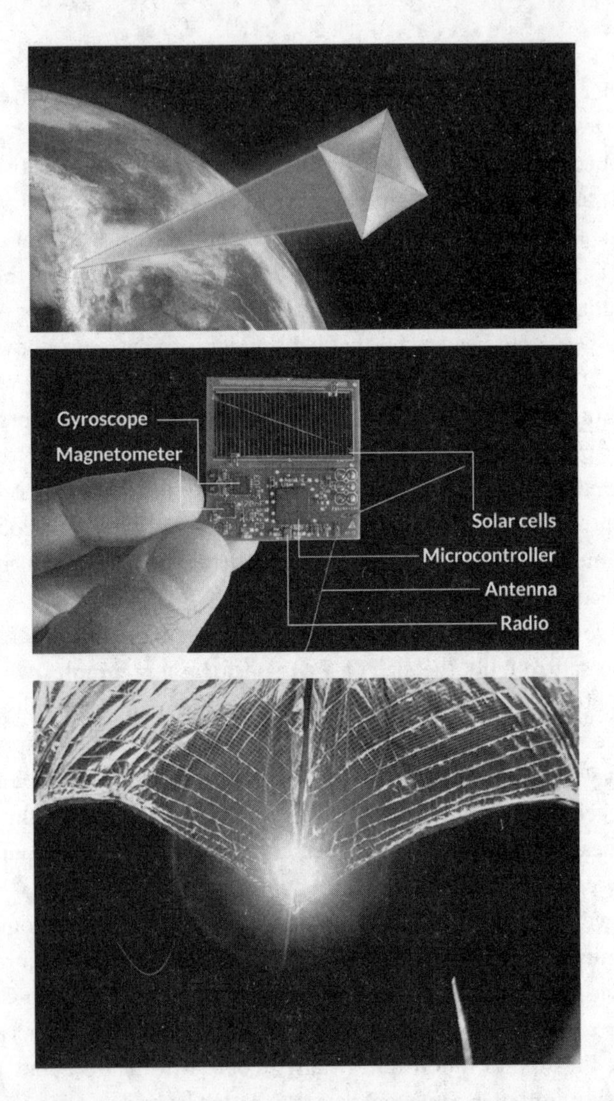

Top: Artist's impression of Starshot, a project to push a lightsail from Earth with a powerful laser beam. Middle: An example of the lightweight electronic device (StarChip) that could be attached, along with a camera, to the sail. Bottom: Photograph of the solar LightSail 2 deployed by the Planetary Society on July 23, 2019, with the Sun showing through the thirty-two-square-meter sail. Breakthrough Starshot/A. Loeb *(top and middle images)* and Planetary Society *(bottom image)*

Just a few months into this endeavor, in August 2015, I coauthored a paper with my postdoc James Guillochon on lightsails. In it, we mused that since humans had been able to come up with this technology, other intelligent life might too. Building on that hypothesis, we argued that scientists should seek out the sort of microwave beams that extraterrestrial life might use to send such craft of their own among the stars.

When our article was published in the *Astrophysical Journal* in October 2015, there had not yet been a formal announcement of the Starshot Initiative, and James and I had broached only one consequence of my team's exploration of the possible solutions to the Starshot challenge. But it was rewarding to have a published article arise out of our earliest evaluations of Yuri's proposal.

The paper's publication also had an unanticipated consequence: the media paid attention. The media's interest hadn't been one of the goals of my research, which followed the same principles that had guided my previous hypotheses on the nature of dark matter, the first stars, and black holes. In retrospect, I realized this unexpected media attention was a sign of things to come.

• • •

Back at Harvard, we continued to work, and six months after I met Yuri and Pete Worden, I received the expected phone call from Pete. He and Yuri wanted me to report my team's findings. They wanted me to do this at Milner's California home. And they wanted me to do it in two weeks.

I had been ambitious when I asked for a mere six months to draw up a plausible plan for reaching the nearest star in around twenty years. Now I had to quickly sum it all up in a way that would convince a small panel to fund it. And the board would be rounded out by Stephen Hawking, by then the most renowned living theoretical astrophysicist. He was not the only scientific luminary who would be reviewing what I submitted. Freeman Dyson corresponded with me routinely about my research, and he had started showing interest in Starshot as well.

When Pete reached me, I was on vacation and just on my way out the door of my hotel room to visit a peaceful and secluded goat farm in the Negev region of southern Israel where my wife wanted to spend the weekend. So the following morning I prepared a presentation while sitting outside with my back against the wall of the office of the goat farm—the only place with internet connectivity.

For me, it was ideal. The weather was cool and dry. I looked out on the goats that had been born the previous day. It was all very familiar, reminding me of the farm I had grown up on with my two sisters, Reli and Shoshi. Among my tasks had been collecting eggs and helping to corral the farm's just-born chicks when they escaped their cages. It was in this familiar setting that I typed out my plans for humanity's first interstellar probe using lightsail technology.

Two weeks later, I visited Milner's home in Palo Alto and announced that we had a plan that fulfilled the mandates he had set out. Within our lifetimes, it was technologically feasible to send a craft to Proxima Centauri.

Yuri was pleased and excited, as was Pete. After several months of extensive discussions, they decided to publicly announce the Starshot Initiative at the observatory atop One World Trade Center in New York City on April 12, 2016. This was Yuri's Night, a commemoration of the launch of the first human into space, Yuri Gagarin, on April 12, 1961. There, on the stage with Yuri Milner, I was joined by Freeman Dyson and Stephen Hawking. The historic vision presented by the panel was recorded by TV crews and broadcast all over the world. The following morning, my wife brought our car in for an oil change, and the mechanic, who was used to seeing me perform this chore, asked where I was. Ofrit mentioned that I could not do it because of the announcement, and he replied: "This project is amazing; I read all the news reports about it." The vision of visiting another star within our lifetimes captivated the public's imagination in a way reminiscent of the Apollo 11 moon landing.

Just seventeen months after that presentation, 'Oumuamua was discovered by the telescopes of Pan-STARRS.

• • •

Let us pause here briefly to recap the evidence that emerged about 'Oumuamua in the weeks immediately following its discovery. It was a small, oddly shaped, shiny object that deviated from an orbit shaped by the Sun's gravity without showing any discernible cometary tail (caused by the outgassing of a comet's ice turning to steam by friction and the warmth of the Sun) despite a deep search for it with the Spitzer Space Telescope and other detectors.

These are certainties, and they allow us to declare confidently that the first three of 'Oumuamua's identified anomalies—its unusual orbit without a tail, its extreme shape, and its luminosity—make it statistically different, by a large margin, from all other objects cataloged by humanity. To put this distinctiveness in statistical terms, we can conservatively state that based on its extra push and its lack of a cometary tail, 'Oumuamua is a one-in-a-few-hundred object. Based on its shape, it is, also conservatively, a one-in-a-few-hundred object. And based on its reflectivity, it is (again, conservatively) at least a one-in-ten object. When we multiply those three anomalous qualities, we can appreciate how much of an outlier 'Oumuamua is. It is now a one-in-a-million object.

Those three traits—orbit, shape, and reflectivity—do not exhaust 'Oumuamua's weirdness, as we know. Alone, however, these three traits *do* clearly defy the understandable but naive expectation that our first interstellar visitor would resemble the rocky asteroids and icy comets that we know have passed through our solar system.

Yet even as these anomalies mounted, most scientists clung to the most familiar explanation: that 'Oumuamua had to be a naturally occurring object, an asteroid or comet. Most scientists—but not all. Even our community, you see, has its anomalies.

With my work on the Starshot Initiative fresh in my mind, I, for one, was finding myself pulled toward a different hypothesis.

5

The Lightsail Hypothesis

IN EARLY SEPTEMBER 2018, JUST ABOUT A YEAR AFter 'Oumuamua passed overhead, I wrote an essay for *Scientific American* on what the search for relics of alien civilizations — specifically, dead civilizations — might entail. Based on Kepler satellite data, I argued, we knew that about a quarter of all stars hosted habitable Earth-scale planets. Even if only a small fraction of all habitable Earths led to technological civilizations like our own during the lifetime of their stars, there might be plenty of relics out there in the Milky Way for us to explore.

Some of these habitable worlds, I theorized, might have evidence of previous civilizations, anything from atmospheric or geologic traces to abandoned mega-structures. But even more intriguing was the possibility that we would find flying through our solar system technological relics with no detectable functionality — for example, pieces of equipment that had lost power over the millions of years of their travel and become space junk.

I went on to note that it was entirely possible we had already found one such technological relic. Noting the discovery of 'Oumuamua the previous fall, I summarized the anomalous evidence we had accumulated about it and then posed a rhetorical question: Given its deviation from its expected orbit and its other peculiarities, "might 'Oumuamua have been an artificial engine?"

Like my idea of eavesdropping on alien civilizations, it was just a passing thought. And I might have been content to let it stay that way if I could have gotten the StarChips out of my head.

• • •

About that time, a new postdoctoral fellow, Shmuel Bialy, arrived at Harvard's Institute for Theory and Computation, where I serve as director. I proposed to him that we collaborate on a paper explaining 'Oumuamua's excess acceleration through radiation from sunlight. Because of my previous work on lightsails in conceptualizing the Starshot Initiative, I was familiar with the scientific constraints and possibilities that interstellar travel by lightsail technology presented. The relevant formulas were all fresh in my mind and ready to be applied to possibly explain the peculiar force exerted by sunlight on 'Oumuamua. To be clear, my attitude at the time was simply *That might work*. The astronomical world had been presented with an exciting discovery, an interstellar object, about which we had collected a trove of confounding data. We confronted facts that were hard to match to a hypothesis that accounted for all of them. When I proposed that Bialy and I explain 'Oumuamua's deviation by way of sunlight, I was following the same scientific tenet I had always followed—a hypothesis that satisfied all the data ought to be considered.

Bialy checked the numbers and his excitement grew; the idea I had proposed looked like a viable possibility. This led to a new question: What would we need to assume about 'Oumuamua's size and composition to explain its deviation? The key question was how thin 'Oumuamua had to be in order to possess the extreme

Artist's impression of 'Oumuamua as a lightsail (left) alongside a conventional rendering of the object as an oblong, cigar-shaped rock (right).
Mark Garlick for Tähdet ja avaruus/Science Photo Library

area-to-volume ratio that accounted for its excess acceleration. We determined that 'Oumuamua needed to be less than a millimeter thick for the force of sunlight to be effective.

The implication of this was obvious: Nature had shown no ability to produce anything like the size and composition of what our assumptions suggested, so something or someone must have built such a lightsail. 'Oumuamua must have been designed, built, and launched by an extraterrestrial intelligence.

It is an exotic hypothesis, without question. But it is no less exotic than other hypotheses that have been proposed to explain the outlier characteristics of 'Oumuamua. Nature has shown no propensity to produce pure-hydrogen comets or fluffy clouds of materials that are both more rarefied than air and structurally cohesive. The extraordinary nature of our conclusion rested almost entirely on the presumption that it wasn't a naturally occurring object.

The lightsail inference may seem outlandish, but getting to it did not require any wild leaps. Shmuel and I went down a logical path. We followed the evidence, and, in the grand tradition of the detective work of science, we hewed closely to a maxim of Sherlock Holmes: "When you have eliminated the impossible, whatever remains, *however improbable,* must be the truth." Hence our hypothesis: 'Oumuamua was artificial.

We laid out these ideas in a paper titled "Could Solar Radiation Pressure Explain 'Oumuamua's Peculiar Acceleration?" In it, we confronted a range of other questions about 'Oumuamua. We described the likely damage it would sustain as it soared across the universe, for instance, from colliding with space dust or from the continuous strain of centrifugal force caused by its rotation. We discussed the impact that such damage might have on the object's mass and speed and found it to be minimal. Laying out equation after equation, we drew conclusions from the available data about the object's thickness and mass, which dictated its surface-to-volume ratio. And then, at the end of the paper, we put forth our hypothesis: "If radiation pressure is the accelerating force," we wrote, "then 'Oumuamua represents a new class of thin interstellar material, either produced naturally . . . or is of an artificial origin.

"Considering an artificial origin," we continued, "one possibility is that 'Oumuamua is a lightsail, floating in interstellar space as debris from advanced technological equipment."

We submitted the paper to the prestigious scientific journal the *Astrophysical Journal Letters,* which specializes in timely and high-impact papers, in late October 2018. Our intention was to engage the attention of our fellow scientists who we knew were weighing hypotheses against the evidence. This, too, should be considered. In that spirit, we also posted the manuscript before it had been peer-reviewed on the online preprint site arXiv.org. Science journalists regularly scour arXiv for stories, and in short order two of them found our study and raced to report our hypothesis. Their pieces went viral, and by November 6, 2018, roughly a year after 'Oumuamua was discovered, everything blew up.

• • •

Within hours after the first media reports appeared, I was surrounded by cameras. While most of America was filing off to vote in the hotly contested midterm elections, four television crews crowded into my office on Garden Street in Cambridge, Massachusetts. I tried to field their questions while simultaneously responding to a steady stream of phone calls and e-mails from newspaper reporters.

I had some experience with the popular press because of earlier papers I had written on an array of subjects, but this level of attention was new to me, and a bit overwhelming. It didn't help that I was mentally preparing to depart for Berlin to give a public lecture, long in the works, at the Falling Walls Conference—an appropriately named gathering dedicated to celebrating breakthroughs that broaden society's interest in the latest advances in science and technology.

I raced home and grabbed my suitcase but didn't make it back out to the car before another film crew arrived, having tracked down my home address. As I stood at my front door, the reporter asked, "Do you believe there are alien civilizations out there?"

"A quarter of all stars host a planet the size and surface temperature of the Earth," I said into the camera. "It would be arrogant to think we are alone."

By the time I deplaned in Berlin, members of the international media were responding much as the American media had. All of this before our paper was even published.

Given the media's attention and the need for us to present more of the factual basis for the hypothesis, the *Astrophysical Journal Letters* published our paper on November 12. They accepted it for publication just three days after I submitted it, the fastest turnaround time in my entire scientific career.

I was grateful for the paper's publication; it meant that an ever-widening circle of scientists was considering our hypothesis to explain the evidence left by 'Oumuamua. But I was under no illusion that any appreciable part of the scholarly field would approach the theory that 'Oumuamua had originated in an extraterrestrial civilization as just one exotic hypothesis among many. I presumed a majority would be reluctant to consider the idea and that some scientists would even be hostile. I was well aware of the prevailing suspicion regarding any argument that kept company with SETI scientists' thinking.

The outpouring of popular interest—which only grew with the release of our paper—seemed ironic, too, when I considered the relative tameness of the hypothesis. Just a year earlier, following reports of an anomaly having to do with hydrogen atoms (which had been found to be colder than expected in the early universe), I published a paper with another Harvard postdoc, Julian Munoz, showing that if the dark matter was made of particles that possessed a tiny electric charge, they would cool the cosmic hydrogen and account for the reported anomaly. Even though this hypothesis was published in *Nature* and even though it was far more speculative than my and Bialy's hypothesis about 'Oumuamua being alien technology, it garnered much less attention.

To be clear, although I made myself as available as my commitments allowed, I neither sought the limelight nor particularly enjoyed it. In the past when I had set out to draw attention to my work, as with the Starshot Initiative, I was grateful when even a few members of the press responded. And while I had undergone extensive professional training in various fields throughout my life, no one, especially me, had thought to include media training. In hindsight, maybe someone should have. Astronomy and astrophysics are fields that frequently require substantial commitments of time and money, and harnessing the public's awareness of what is possible and what might be necessary cannot be an afterthought.

• • •

To say that my raising the possibility that ʻOumuamua was artificial technology met with disapproval is putting the matter mildly. To be sure, the popular media was delighted, and the broader public was fascinated. But my fellow scientists were, shall we say, more circumspect.

In July 2019 the ʻOumuamua Team of the International Space Science Institute (ISSI) published their unambiguous conclusion in *Nature Astronomy:* "we find no compelling evidence to favor an alien explanation for ʻOumuamua." The immediately preceding paragraphs declared that the extraterrestrial-technology theory that Bialy and I had put forward was provocative but baseless. Yet the article also ends with a list of unanswered ʻOumuamua anomalies, what the authors termed "open questions." They also acknowledged that only after the advanced telescope at the Vera C. Rubin Observatory in Chile was fully operational might we have sufficient data to determine "how common — or rare — the properties of ʻOumuamua are."

It was never my intent to become what the science journalist Michelle Starr labeled me: "Harvard astrophysics *enfant terrible.*" My attitude toward anomalies remains what it has been since my first day of grade school — quizzical and questioning; I pause long enough to wonder what might follow if I pursue one course of action over another. When Starr asked Matthew Knight, an astronomer at the University of Maryland and one of the scientists on the ISSI ʻOumuamua Team, to sum up the team's findings, he declared, "We have never seen anything like ʻOumuamua in our solar system. It's really a mystery still," and then added, "but our preference is to stick with analogues we know."

Fair enough. But what happens when we start from the mystery end of the trench rather than the familiar-analogues end? What questions arise and what new avenues for pursuing answers suggest themselves when we entertain possibilities that are counter to our governing presumptions but align with the data we have?

This is not an idle question; the data we have forces us to entertain exceptionally rare explanations for it. Several other mainstream astronomers, not members of the above groupthink, analyzed 'Oumuamua's data carefully and found that only highly exotic explanations could account for the object's anomalous behavior. To explain all the known facts, they were forced to imagine 'Oumuamua was a fluffy object composed of material a hundred times more rarefied than air or that it was a comet composed of solid hydrogen ice.

Scientists had to offer up "never seen before" options to explain Oumuamua's proven peculiarities. Of all the many asteroids and comets that we have cataloged, none have had such peculiarities. If these scientific-mainstream explanations for 'Oumuamua were deemed valid enough for thoughtful consideration, the hypothesis that it was extraterrestrial technology, similarly a "never seen before" possibility, deserves no less.

The questions that the lightsail hypothesis raises, moreover, are intriguing. If we suppose that 'Oumuamua is an exceedingly rare comet composed of frozen pure hydrogen, most of our questions dead-end. The same is true if we imagine it as a fluffy cloud of dust with sufficient internal integrity to hold together but still lightweight enough to be pushed by sunlight. In both instances, we can marvel, but that's about all we can do. Statistical rarities belong on the shelves of a curiosity cabinet; they shouldn't give rise to new branches of scientific inquiry. But if we acknowledge that 'Oumuamua is plausibly of extraterrestrial-technology origin and approach that hypothesis with scientific curiosity, whole new vistas of exploration for evidence and discovery open before us.

After the media had gotten over the initial shock that the chair of Harvard's astronomy department and his postdoc were postulating that 'Oumuamua was a relic of extraterrestrial technology, I was accused of seeing lightsails wherever I looked. After all, my involvement in the Starshot Initiative had been announced just two years earlier and our stated goal was to send electronic chips to the nearest star by harnessing the power of lightsail technology.

The interviewer for the German newspaper *Der Spiegel* put it with admirable bluntness: "According to a proverb, whoever has only a hammer will see nothing but nails."

I replied that, yes, like everyone's, my imagination was guided by what I knew, and, yes, like everyone's, my ideas were influenced by what I was working on. But I should have added that the problem with the proverb was that it focused attention on the hammer rather than on the person wielding it. Not only do skilled carpenters most definitely *not* see nails everywhere, but they are trained to differentiate among those they do observe.

6

Seashells and Buoys

One of my favorite activities is walking along the beach searching for seashells worth collecting. It is a pastime I indulge in while I'm on vacation, when I get to pick a beloved stretch of sand to stroll and study and I have the free time to do it. My daughters often join me as we examine what has been swept ashore. Over the years we have amassed a nice collection of delicately attached bivalves, rounded cowries, and curled triton and murex shells. A few of our shells are pristine, but far more are worn down and partially disintegrated, their tiny pieces now part of the white sand we've walked on.

Sometimes while searching for shells, we find a piece of sea glass—a shard from a discarded bottle that, after years of being tossed and tumbled in the ocean, has been made smooth. Under these conditions, even industrial products can be things of beauty.

Occasionally on our shell-hunting expeditions, we find other, less beautiful man-made objects—a plastic bottle, say, or an old grocery bag. But such discoveries are relatively rare, and that rarity is easily explained: we try to vacation near places where we are less likely to encounter trash.

If our family wished, we could travel to beaches where encountering trash is a certainty. Sadly, our planet has a growing number of them. For instance, Hawaii's Kamilo Beach; once beautiful, it is now known as "Plastic Beach" due to the extent of garbage that accumulates there. Its condition is sad but not all that surprising, given that the Great Pacific Garbage Patch — by some estimations, the largest of the world's five offshore "plastic-accumulation zones" — sits between California and Hawaii. And the existence of five such patches is also not surprising, given that humanity puts about eight million metric tons of plastic in the oceans each year.

The more there is of something, the more likely it is you will encounter it. This truism applies equally to seashells and plastic bottles — and to the two potential explanations for 'Oumuamua I have described so far. Either it is a naturally occurring seashell or it is a piece of manufactured material, junk or otherwise.

Viewing both possibilities through a lens made of beach glass, we can see why identifying the right one is so important — and what implications it will have for both science and our own civilization.

• • •

Let's suppose that, rather than a plastic bottle, 'Oumuamua was a seashell. An exotic seashell, certainly, but still a naturally occurring seashell.

This line of reasoning has attracted the vast majority of scientists who have considered 'Oumuamua's anomalies. It comes undone almost immediately, however, when we ask how many interstellar seashells must there be for our solar system to have randomly come across one.

No one is surprised to encounter an intact seashell while walking along the beach. The sea creatures that produce shells are vast in number, and even given the size of the world's oceans, there are more than enough of them to make it commonplace to find shells worth collecting. Indeed, if we wished, we could estimate the likelihood of encountering not just *a* seashell but a specific *kind* of seashell along a given stretch of beach. Knowing a little about the number of quahog clams in the waters around Cape Cod, for example, would allow us to predict how often we could expect to discover one on the beaches around Provincetown. The same would go for a conch shell on a Florida beach.

If 'Oumuamua is a naturally occurring asteroid or comet, we can pose the following question: How many interstellar rocks need to be in the universe for human beings to regularly encounter them in our solar system? If interstellar space has a large population of asteroids and comets, just like the familiar family of those bound to the Sun, it wouldn't be surprising for us to see them. After all, as I've noted, the more there is of a thing, the more likely you are to encounter it. If interstellar space has a *small* population of such rocks, however, it would be more surprising to encounter them in our solar system.

Interstellar space is, of course, many magnitudes vaster than the Earth's oceans. This means that for us to regularly encounter them in our solar system, there would have to be a very, very large population of such interstellar objects floating around out there. Such rocks are known to be the building blocks of planetary systems around stars.

Actually, *very, very* doesn't come close to doing this inferred population justice. To account for a population of rocks as large as the discovery of 'Oumuamua implies is out there requires that each star in the Milky Way eject 10^{15} such objects from the rocky material around it during its lifetime. To get a sense of the size of that number—a quadrillion—grab a piece of paper and write down a 1 followed by fifteen 0s. It's not quite the size of the number of habitable planets in the observable universe (10^{21}), but still, it represents a lot of objects coming from each and every star in our galaxy. Planetary systems around stars are the regions where large solid objects are likely to form.

Our own Sun has not been nearly so profligate with its planetary building blocks. In 2009, nearly a decade before ‘Oumuamua’s discovery, I published a paper with Amaya Moro-Martin and Ed Turner in which we used a dynamical history of our solar system to predict a population of random interstellar objects; it is smaller by two to eight orders of magnitude than the amount needed to explain the discovery of ‘Oumuamua. In other words, the number of predicted interstellar objects our projection came up with was at least one hundred times lower than that needed for the hypothesis that ‘Oumuamua was a random interstellar rock. This by itself does not rule out ‘Oumuamua being a familiar rock, but it does make its discovery in our solar system surprising on statistical grounds.

Put another way, the idea that ‘Oumuamua was a naturally occurring rock implies that the population of random interstellar objects is far greater than what we expect and what our own solar system predicts. So either a great many other stars out there are very different from the one nurturing us, or there is something else going on.

• • •

In 2018, a small group of scientists revisited the question of the abundance of ‘Oumuamua-like rocks in interstellar space. While studying the ability of Pan-STARRS to detect objects similar to ‘Oumuamua, they reached some general conclusions. Among them was the broadly agreed-upon insight that “many aspects of ‘Oumuamua are both intriguing and troubling.” But they also found that the number per unit volume of interstellar material necessary to have ‘Oumuamua be a random rock requires “mass ejection rates” that far exceed expectations, up to a quadrillion (10^{15}) ‘Oumuamua-size objects per star, yielding roughly one object for every interstellar volume whose circumference is carved by the orbit of the Earth around the Sun. In two follow-up papers, my former collaborator Amaya Moro-Martin showed that the natural abundance of 'Oumuamua-like objects on random orbits falls short of the required value by several orders of magnitude, even if every planetary system ejected all of the expected solid material in it.

These conclusions complicate this echo of our 2009 finding in interesting ways. One complication has to do with the origins of interstellar material, which falls into two broad categories: dry rocky material (which has little to no ice) and icy rocky material.

Dry rocks are created primarily during planet formation. This happens as a result of dust particles sticking together and growing in size to planetesimals, which eventually combine to make planets. But the first of the above studies concluded that the number density needed for the random-rock hypothesis to explain a naturally occurring 'Oumuamua "cannot arise from the ejection of inner solar system material during planet formation." Not enough material is ejected during planet formation to get us to the necessary density.

To get to the needed density of naturally occurring objects, these scientists had to posit an additional source of interstellar objects like 'Oumuamua. And for that, they turned to material ejected from stars' Oort clouds—the shells of icy objects at these systems' outermost edges. When a star reaches the end of its lifetime, its gravitational hold on its Oort cloud bodies weakens, and they are released. But even if all dying stars contribute their Oort cloud debris to interstellar space, Amaya Moro-Martin found in her second paper, they do not provide enough material to achieve the needed density.

The challenge that the "natural origin" explanation for 'Oumuamua confronts is the need for a sufficient amount of interstellar material. The rough analogy of seashells helps; you need a great many seashells in the sea to make discovering an intact one on a beach probable. The same applies to a naturally occurring 'Oumuamua arriving in our solar system. For it to be a randomly encountered object, we need a lot of such objects in the universe, and to get at that density, we need objects released from both planet formation and Oort clouds.

Of course, we have already established that 'Oumuamua wasn't icy. (No outgassing, no ice.) So a naturally occurring 'Oumuamua is very unlikely to have come from an Oort cloud.

In short, if 'Oumuamua was a natural object, it had to have been generated by planet formation. It also has to belong to an unknown class of objects generated by planet formation whose size, shape, and composition make them deviate from a path shaped solely by our Sun's gravity without any visible outgassing.

At the time of this writing, we know of no other object that fits the second set of criteria. But we know of at least one that fits the first.

• • •

Not long after the discovery of 'Oumuamua, we encountered our second interstellar object. By the time you read this book, we may well have found others.

This second interstellar object is named 2I/Borisov, after Gennadiy Borisov, a Russian engineer and amateur astronomer who on August 30, 2019, using a sixty-five-centimeter telescope of his own construction, identified the object in the skies above Crimea. And it was Borisov who first ascertained that its trajectory was hyperbolic. Just as had been true for 'Oumuamua, 2I/Borisov was moving too fast to be gravitationally bound to the Sun. And so, just like 'Oumuamua, 2I/Borisov had come from outside our solar system and was on a path that would send it through and beyond our solar system.

But otherwise, 2I/Borisov was unremarkable. It was an interstellar comet, without question, and for this reason it was distinctive; any interstellar object is a rarity. But its distinctiveness ended there. Its coma and outgassing resembled our solar system's comets' in all characteristics; 2I/Borisov was icy and decidedly not exotic.

The point is that the discovery of 2I/Borisov did not help us move toward a naturalistic explanation for the exotic 'Oumuamua. If anything, it did the opposite, by underscoring how special 'Oumuamua truly was. When I met my wife and realized how special she was, I married her. The many people I have encountered since then do not take away from her unique qualities; they only add to my sense of wonder at how rare she is.

'Oumuamua and 2I/Borisov were both interstellar interlopers in our solar system, but other than that, they were decidedly different from each other, for among 2I/Borisov's list of ordinary features, it was from an unexceptional origin in space-time.

'Oumuamua wasn't. Indeed, its origin in velocity-position space is another of its marked peculiarities—and another piece of evidence supporting an unusual origin. It is also another clue that can help us to unravel the mystery of what 'Oumuamua was and what it was doing out there in the void of interstellar space.

To understand this requires an understanding of velocity-position space. This can be a bit tricky to wrap your mind around, but it boils down to appreciating that the position an object holds in space is defined not just by where it is vis-à-vis everything around it but also by what its velocity is vis-à-vis the velocity of everything around it. Imagine a very busy, very wide multilane interstate filled with thousands of cars. All of them are traveling at slightly different rates; some are passing cars, others are falling back, some are well below the speed limit, and others are greatly exceeding it.

If you averaged these cars' movements, you would find a few cars that were, relative to all the others, "at rest." These cars would not be pulling ahead or falling back relative to the rest of the pack. Amid all that motion, those cars would be comparatively still.

The same applies to the stars. All the stars in the vicinity of the Sun are moving relative to one another. The average of their mo-

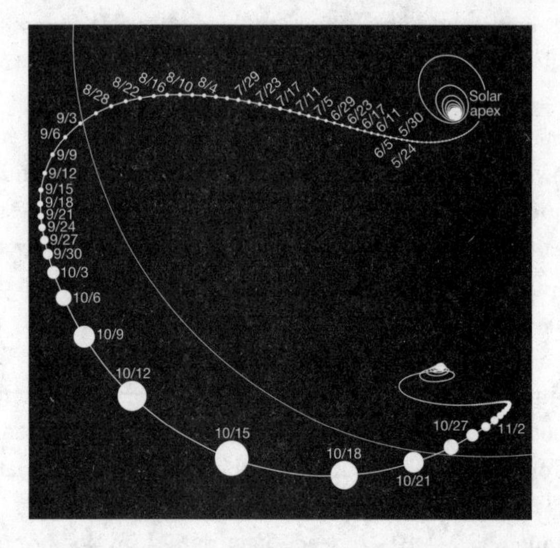

Sky path of 'Oumuamua as seen from Earth, with the stages of the object's progress labeled by date. The relative size of each circle gives a schematic sense of the changing distance of 'Oumuamua along its apparent trajectory. Also shown is the direction of motion of the Sun in the local standard of rest, or LSR (the direction just left of the label "solar apex"). The fact that the object starts from that direction and gets close to us implies that it was initially in the LSR. Between September 2 and October 22, 2017, 'Oumuamua's trajectory moved from the local standard of rest to south of the ecliptic plane of the solar system (marked by the thin line). Image by Mapping Specialists, Ltd. adapted from Tom Ruen (CC BY 4.0)

tions is called the local standard of rest. Amid the motion of all these stars, an object at the local standard of rest, or LSR, is comparatively still. And it is a comparatively rare occurrence.

'Oumuamua was occupying the LSR.

Or at least, it was before it accelerated. Around the time that it encountered our Sun, it went from sitting still—relative to the average motion of the stars in our neighborhood, including our own—to moving away from us. Thanks to the kick it received from the Sun's gravity, it was knocked out of LSR, much as if one of those "still" cars on that multilane highway was violently sideswiped. As a consequence, 'Oumuamua was broken out of LSR and sent on a path along which it would, like a tennis ball hit by a racket, speedily depart our solar system.

'Oumuamua's being at LSR was peculiar. Consider that only one in every five hundred stars is as still within the LSR frame as 'Oumuamua was before it was sideswiped. Our own Sun, for example, is moving at about 45,000 miles per hour relative to this frame, about ten times faster than 'Oumuamua was moving before the Sun kicked it away from the LSR.

What explains an object being at the LSR? What would need to have happened to place an object at such a distinctive velocity in our vicinity? As with all of 'Oumuamua's peculiarities, our answers depend on the assumptions we make concerning its origins.

Let us start with a hypothesis that will be more to the liking of most scientists than my lightsail one: If we suppose that 'Oumuamua was a dry rock, then perhaps the parent star from which it was ejected was one of those one-in-every-five-hundred stars in the LSR frame.

Could this explain the fact that 'Oumuamua, too, is at LSR? Yes, perhaps—if its departure from its home system was exceedingly gentle. Understanding why requires only common sense: An object violently ejected from a star system at LSR would be kicked into a different frame of reference. Only an object gently ejected from such a home system would retain the frame of its parent.

At the grave risk of straining the analogy, let us return to the multilane highway. Envision a motorcycle that is among the few vehicles that are "at rest" in comparison to the cars and trucks traveling around them. Now envision that motorcycle with a sidecar attached only by a single, well-greased pin. Immediately after that pin is gently removed, the sidecar, like the motorcycle, would remain at comparative rest. And if—and here is where we really strain the analogy—the highway was frictionless, motorcycle and sidecar would retain their position and velocity in comparison to all the traffic around them. Similarly, if a planet at LSR detaches a piece of itself gently, the detached piece would retain the planet's position at LSR.

A gentle departure from a parent star is possible but statistically unlikely. Pieces of planets do not detach easily, and the sort of event that would detach a piece of a planet is rarely anything you could describe as gentle. A strike to a planet at LSR that resulted in that planet ejecting an object that remained at LSR would be akin to an extremely careful strike undertaken with the planetary precision of a feather. The probability of it happening is estimated at 0.2 percent.

It's also possible that 'Oumuamua came from one of the 99.8 percent of stars that are in significant motion relative to the LSR. But for that to happen, the ejection maneuver would have had to be a haymaker rather than a nudge, and a precise one at that. The kick that would eject an object from a star system *not* at LSR in a way that resulted in the ejected object *being* at LSR means a blow exactly equal and opposite to the velocity of the parent star. The blow would need to perfectly cancel out the home system's movement to produce an object at the LSR. Imagine the challenge of a surgeon attempting a delicate operation with a crude instrument like a hammer.

Either possibility, feather or hammer, renders exceedingly unlikely this hypothesis about a naturally occurring 'Oumuamua being ejected from a home system at LSR.

That leaves us with a third, only slightly more plausible hypothesis.

An object expelled from a home system at LSR could itself remain at LSR if its expulsion occurred on the extreme outskirts of the home system. There, of course, the gravitational hold of the parent star is far weaker. Indeed, Oort cloud–like outer shells are where most interstellar asteroids and comets that manage to break free of their birth systems probably come from. The more tenuous the gravitational hold of the birth star, regardless of whether it is at

LSR or not, the easier it is for some of the debris in its outermost shell to be pulled away by another source of gravity.

Our own solar system's trillion-comet Oort cloud is a case in point. Its icy shell is 100,000 astronomical units (AUs) distant from the Sun (each AU is the distance from the Earth to the Sun, or roughly 93 million miles). The Sun's gravitational hold on the material making up the Oort cloud is far weaker than its hold on, say, the Earth. Way out there, a mild nudge of less than 2,200 miles per hour—what an encounter with a passing star might provide, for example—could be enough to send an object on its interstellar way.

So, if 'Oumuamua originated from an icy Oort cloud–like shell around a host system in the LSR, that could explain the object's velocity. But this doesn't explain 'Oumuamua being a dry rock.

No matter how you look at it, the dynamical origin of 'Oumuamua—that it was at LSR before encountering our solar system—is extremely rare, and it's all the rarer given our need for a naturally occurring object dry enough to produce no visible outgassing when it deviates from a trajectory explicable by the gravitational force of the Sun alone.

Which brings us to our second hypothesis, that 'Oumuamua was a manufactured object specifically designed to be at the local standard of rest. Perhaps long, long ago, 'Oumuamua was not junk but extraterrestrial technological equipment built for a distinct purpose.

Perhaps it was something closer in intent to a buoy.

• • •

We think of 'Oumuamua as hurtling toward us, but it can be more instructive to view things from 'Oumuamua's vantage point. From that object's perspective, it was at rest and our solar system slammed into it. Or, in a way that works both metaphorically and, maybe, literally, perhaps 'Oumuamua was like a buoy resting in the expanse of the universe, and our solar system was like a ship that ran into it at high speed.

The hypothesis that intelligent extraterrestrials designed 'Oumuamua to be at LSR raises the obvious question: Why would they bother? I can imagine any number of reasons. Perhaps they wanted to set up the interstellar equivalent of a stop sign. Or maybe it was more like a lighthouse—or, more simply, a signpost or navigation marker. A vast network of such buoys could act as a communication grid. Or it could be used as a trip wire, an alert system triggered when one of them was knocked out of LSR. In that spirit, perhaps its creators wanted to disguise its—and their—spatial origins. Putting an object at LSR effectively camouflages who put it there. Why? Because math and a little knowledge of an object's trajectory is sufficient to trace that object's origins back to a launchpad; doing that is one of the primary purposes of the North American Aerospace Defense Command (NORAD). Consider as well that any intelligence with a grasp of math and a good map of the universe could trace back to Earth any of the interstellar ships we've launched from our planet's surface.

That all of these analogies are terrestrial isn't just a reflection of the fact that so is this book's author. That human civilization has built buoys, grids of communication satellites, and early-warning detection systems tells us that it is probable other civilizations might do the same thing. What is more, these conjectures are plausible for the simple reason that any one of them is an outcome that humanity could engineer, build, and launch if we wished to. Our reasons needn't even be interstellar. Just for example, if India had put such an object into space, the scientists at NASA might wonder why, but they wouldn't wonder how there came to be a small, flat, luminous object marked with the Indian Space Research Organisation's distinctive logo at LSR.

The hurdle to accepting that this answer applies to 'Oumuamua is, of course, that it requires us to accept that 'Oumuamua is of extraterrestrial origin. And the hurdle to that is that we must take seriously the possibility that we're not the only intelligence in the universe.

• • •

A buoy. A grid of pods for communication. Signposts that an extraterrestrial civilization could navigate by. Launch bases for probes. Other intelligent living organisms' defunct technology or discarded technological trash. These all are plausible explanations for the 'Oumuamua mystery—plausible because here on Earth, humanity is already doing these things, albeit on a far more limited scale, and we would certainly consider replicating them if and when we explore out into interstellar space.

What renders these hypotheses implausible is the inability to conjecture an extraterrestrial intelligence. Foreclose that possibility, and you moot all such explanations. Refuse to look through the telescope and it matters little if it does or doesn't show any compelling evidence. Perhaps it is the shadow of science fiction stories or just a flaw in some people's ability to broaden the scope of hypotheses they will entertain, but entertaining an explanation that posits an extraterrestrial civilization is much like presenting to the skeptical a telescope they outright refuse to look through.

The best antidote to such recalcitrance, I have found, is to think for yourself. If any of these ideas seem feverish or over the top or detached from reality, just remind yourself of the evidence before you.

The data we confront tells us that 'Oumuamua was a luminous, thin disk at the LSR, and when it encountered the gravitational pull of the Sun, it deviated from a trajectory explicable by gravity alone, and it did so without visible outgassing or disintegration.

These data points can be summed up as follows: 'Oumuamua was statistically a wild outlier.

Using very conservative probabilities, based on its shape, rotation, and luminosity alone, a cometary ʻOumuamua would be a one-in-a-million naturally occurring object. Attempt to explain its composition so that we can explain its deviation beyond solar gravity by outgassing that was invisible to our instruments and you still have an object that is as rare as one in thousands.

But that's not all. It's also very odd, remember, that ʻOumuamua's spin rate didn't change. Maybe just one in every thousand comets keeps a steady spin despite the significant mass loss implied by ʻOumuamua's nongravitational acceleration. If ʻOumuamua was one of those rare comets, we're now talking about a one-in-a-billion object.

Then there is its lack of jerks. If there was naturally occurring outgassing and disintegration that was somehow invisible to our instruments, those putative jets propelling ʻOumuamua would have had to perfectly cancel each other out. If that's another one-in-a-thousand coincidence, ʻOumuamua is now one in a trillion.

And we still have to take into account ʻOumuamua's origin in velocity-position space, the fact that it was at LSR. Recall that an LSR birth star is a 0.2 percent probability, so now the odds that ʻOumuamua is just a random comet are approaching one in a quadrillion.

These numbers strain credulity and beg for an alternative explanation. They were what led me to suggest to Shmuel Bialy that we search for another, more plausible hypothesis. And we could come up with just one that made sense for the nongravitational acceleration: ʻOumuamua's weirdly steady push was provided by sunlight.

This was fully consistent with an important clue. The excess force acting on ʻOumuamua that caused it to deviate, observers noted, seemingly declined inversely with the square of its distance from the Sun, as one would expect if that force was provided by reflected sunlight.

But solar radiation pressure isn't very powerful. If it were indeed responsible, we calculated, then 'Oumuamua had to be less than a millimeter thick and at least twenty meters wide. (The diameter depends on the object's reflectivity, which is unknown. If 'Oumuamua were a perfect reflector, bouncing back 100 percent of the sunlight that hit it, in this super-thin scenario it would be twenty meters across.)

There is nothing in *nature* with those dimensions, as far as we can tell, and no known natural process that can produce them. But of course, humanity has built something that fits the bill, and we've even launched it into space: a lightsail.

We arrived at this hypothesis through logic and evidence—in short, by sticking to the facts. But if we take the hypothesis seriously, it allows us to ponder new, incredible questions about how 'Oumuamua appeared in our universe and where it came from. It even, as I will explain, gives us the opportunity to ask whether we might someday meet the creators of this mysterious visitor.

The lightsail hypothesis opens up a world of possibilities—unlike the comet hypothesis, which closes them off. The fact that scientific consensus strongly favors the more conservative and restrictive of these two possibilities says less about the evidence than it does about the practitioners and culture of science itself.

7

Learning from Children

ARE WE ALONE? THIS QUESTION IS AMONG THE most fundamental humanity confronts. The moment we have a conclusive answer, negative or positive, is the moment we face profound realizations. Indeed, there are few cosmological questions of equal importance.

To be sure, it would be transformative to learn what preceded the Big Bang, where the matter sucked into a black hole goes, or what theoretical insights finally square relativity with quantum physics. Indeed, I have devoted a significant portion of my life and career to answering the first two of these questions. But would the answers to these questions change our sense of ourselves as significantly as learning that we are just one intelligent species among many—or, conversely, that we are the only conscious intelligence to arise in the universe? I doubt it.

Because I believe this question is so consequential, I find it remarkable how rarely, and how cavalierly, scientists have gone about seeking an answer. That didn't start with the resistance to my lightsail theory; far from it. Scientists' reluctance to read the messages of 'Oumuamua long predates its passing through our solar system.

• • •

The search for extraterrestrial life has never been more than an oddity to the vast majority of scientists; to them, it is a subject worthy of, at best, glancing interest and at worst, outright derision. Few of repute have dedicated their careers to advancing the field, and even at its zenith of academic respectability, in the 1970s, only about a hundred scholars were publicly associated with the SETI Institute. Far more speculative fields of mathematical gymnastics are known to attract bigger communities of physicists.

The more rigorous approach to SETI began in 1959 when two physicists based at Cornell University, Giuseppe Cocconi and Philip Morrison, coauthored a seminal paper titled "Searching for Interstellar Communications." Their article, which was published in the prestigious scientific journal *Nature,* made two simple conjectures. First, that extraterrestrial civilizations as advanced as ours if not more so existed. Second, that such civilizations would most likely broadcast their interstellar communication *We exist* at the radio frequency 1.42 GHz, that "unique, objective standard of frequency, which must be known to every observer in the universe." Cocconi and Morrison were referring to the twenty-one-centimeter wavelength of neutral hydrogen — the very same radio emission that would preoccupy me and other astrophysicists nearly half a century later as we attempted to peer back in time to the cosmic dawn.

The paper was an immediate sensation, heralding the birth of SETI and establishing the rationale for all subsequent searches for extraterrestrial intelligence with its concluding sentence: "The probability of success is difficult to estimate; but if we never search, the chance of success is zero." For me, it echoes a much older thought, one attributed to Heraclitus of Ephesus: "If you do not expect the unexpected, you will not find it."

Cocconi and Morrison's paper also brings to mind the old saw about people with nothing but hammers seeing only nails everywhere they look. The two men were writing a quarter of a century after the birth of radio astronomy, a fact that surely helped them in their effort to "expect the unexpected." As with my and Bialy's lightsail hypothesis, humans seem to be better at seeing the technological signature of alien civilizations after we have developed the technology ourselves.

Cocconi and Morrison's paper immediately inspired the astrophysicist Frank Drake. In 1960 he decided to conduct just the kind of search that Cocconi and Morrison encouraged. Using the National Radio Astronomy Observatory of Green Bank, West Virginia, Drake conducted a search of two nearby sun-like stars, Tau Ceti and Epsilon Eridani. For one hundred and fifty hours, spread out over four months, Drake used the radio telescope to seek a discernible signal indicating intelligence, with no success. The sense of whimsy with which Drake undertook his search for extraterrestrial life is captured in the name he gave the project: Ozma. It derives from a character invented by the novelist L. Frank Baum in his Land of Oz books.

Drake's project became the stuff of broad interest and popular media coverage. That about two hundred hours of observation discovered nothing hardly dimmed public enthusiasm. And on the updraft of that interest, in early November 1961, Drake participated in an informal conference sponsored by the National Academy of Sciences at the National Radio Astronomy Observatory. It was there that he first articulated the Drake equation, which he used to estimate the number of actively communicating extraterrestrial civilizations.

The equation now graces T-shirts, informs the plots of young adult novels, was misused by Gene Roddenberry to bestow a patina of plausibility to the television series *Star Trek,* and has been roundly critiqued and tweaked by scientists ever since. Lost in the dust and noise is the simple appreciation that the equation is nothing more than a heuristic, a shorthand tool to factor out the different terms that affect the success of SETI. Its standard expression is:

$$N = R_{*} \times f_p \times n_e \times f_l \times f_i \times f_c \times L,$$

where the terms are defined as:

N: the number of species in our galaxy that possess the technology necessary for interstellar communication;
R_{*}: the rate of star formation in our galaxy;
f_p: the fraction of stars with planetary systems;
n_e: the number of planets in each system with environmental conditions amenable to life;
f_l: the fraction of planets on which life arises;
f_i: the fraction of planets on which intelligent life arises;
f_c: the fraction of intelligent life that develops sufficiently sophisticated technology to take part in interstellar communication;
L: the duration of time such intelligent life is able to produce detectable signals.

Unlike most equations, Drake's was not designed to be solved. Rather, it was intended to serve as a framework for thinking about how many intelligent civilizations might occupy our universe. It is unlikely we will ever be able to plug in values for all of the variables, let alone determine their output.

While Drake was not alone in formulating a framework for seeking extraterrestrial intelligence—Ronald Bracewell came up with a different approach in 1960 and Sebastian von Hoerner, a German astrophysicist, yet another in 1961—his was the one that has since, for good or ill, become the touchstone for SETI science.

I say for ill because Drake's equation focuses solely on the transmission of communication signals; he limited his aspirations to finding *N* and from it the number of interstellar communications that would establish the existence of extraterrestrial intelligence. This exclusive interest in communication predicts the equation's second limitation, epitomized by its variable *L*, which represents the length of time an intelligent species would be able to produce such signals. Consider, for example, that our species has been producing pollutants that are detectable by certain telescopes for centuries but radio signals for mere decades.

Both *N* and *L* point to a deeper problem with the Drake equation. For all its value as the first systematic effort to identify the variables involved in estimating and thereby directing efforts to find extraterrestrial intelligence, the equation's very formalism was perhaps its biggest limitation. When SETI scientists failed to come up with evidence of alien radio signals, critics were happy to declare the equation—and all of SETI—whimsy through and through.

In 1992, in keeping with the search for *N*, the U.S. government gave 12.25 million dollars to NASA to initiate a radio astronomy program. The very next year, SETI funding was ended. At the time Congress withdrew its support and the funding, one senator, Richard Bryan of Nevada, declared, "Millions have been spent and we have yet to bag a single little green fellow." There are few more succinct statements of the ignorance and flawed assumptions that have hampered humanity's pursuit of the answer to "Are we alone?" The sum spent was paltry, comparatively, and the bar set for evidence of success was absurd.

That said, the early SETI researchers rarely helped their own cause. Their near-exclusive focus on seeking radio and optical signals has set unhelpful presumptions, scientific and popular, as to what such exploration looks like and what sorts of projects merit funding. Only recently have we witnessed growing interest in seeking biosignatures, such as oxygen and methane in the atmosphere and large-scale algal blooms in distant oceans, and technosignatures, such as markers of industrial pollutants in planetary atmospheres and localized heat islands that suggest urban settlements.

In the search for extraterrestrial intelligence, the members of this niche are still finding their footing, and the broader scientific community that ought to be supporting them largely is not. Human science still needs to mature—in regard to SETI as well as other frontiers of our limited imagination.

• • •

I keep a file drawer in my office that is labeled, simply, IDEAS. It holds a single hanging file in which I keep manila folders. Sometimes the file is overstuffed, sometimes less so. In each folder are a few sheets of paper displaying equations. These reflect problems and questions that occur to me and that are worthy of answers. They often keep me company during my walks in the backyard of my home and the nearby woods. At the risk of sounding clichéd, I usually come up with them during my shower. (Recently, after a Dutch film crew visited our shower in an attempt to document my inspiration, my wife bought me a waterproof pen and whiteboard.)

Well before I had a drawer for collecting ideas, let alone undergraduates, graduate students, and postdocs to share these ideas with, I was gathering them. They have served as seeds from which my own research has grown. To date, those seeds have yielded more than seven hundred published papers, six books (including the one in your hand), and a growing number of now-confirmed predictions touching on the birth of stars, the detection of planets beyond our solar system, and the properties of black holes.

This is not to say that I am guided by imagination alone. All my studies reflect an unshakable guiding principle: contact with data. I avoid mathematical speculations, or what I call "theory bubbles." Too often astrophysics can lose itself in theories that float free of any evidence, taking funding and talent with them. There is one reality out there and we are very far from exhausting all of its anomalies.

As I have told generations of students, it is dangerous to wander into work on abstractions that hold little to no promise of feedback from data. I am sure many of them have felt that it is equally dangerous to pursue lines of research or advance conclusions that are against the scientific mainstream. I think that this reaction is not only a shame – it is also dangerous.

While the past several decades have offered much encouragement to the search for extraterrestrial life, I am repeatedly struck by the extent of what remains untried, undertheorized, underfunded, and, among a broad swath of scientists, deemed best left unmentioned. When I describe to my colleagues the reactions of my undergraduates to the two thought experiments that opened this book, many of them chuckle. I think that we should pay closer attention and ask ourselves if there isn't a professional truth hiding in plain sight of the students' responses.

Unlike what trends on social media, scientific progress is measured by how close a proposed idea is to evidence-established truth. That widely accepted fact suggests that physicists would measure their success by how well their ideas align with data rather than by how popular those ideas are. But that is not what we discover when we survey the landscape of theoretical physics. Fashions frequently dictate funding, sometimes despite anything close to a commensurate return on investments.

Despite the absence of experimental evidence, the mathematical ideas of supersymmetry, extra-spatial dimensions, string theory, Hawking radiation, and the multiverse are considered irrefutable and self-evident by the mainstream of theoretical physics. In the words of a prominent physicist at a conference that I attended: "These ideas must be true even without experimental tests to support them, because thousands of physicists believe in them and it is difficult to imagine that such a large community of mathematically gifted scientists could be wrong."

But go beyond the groupthink and look more closely at these ideas. For instance, supersymmetry. This theory, which postulates that all particles have partners, is not as natural as prominent theorists predicted it would be. The latest data from the Large Hadron Collider at CERN did not find any of the evidence expected at the energy scales it probed to support supersymmetry. Other speculative ideas pertaining to the nature of dark matter, dark energy, extra dimensions, and string theory have yet to be even tested.

Imagine that the data suggesting that 'Oumuamua is extraterrestrial technology is stronger than the data suggesting supersymmetry theory is valid. What might follow? Just a bit under five billion dollars was spent to construct the Large Hadron Collider, a particle accelerator built in hopes of attaining confirming evidence of supersymmetry, and running it costs another one billion dollars a year. If the scientific consensus eventually gives up on the theory, it will do so after vast expense and generations of effort. Until we have invested similarly in the search for extraterrestrial intelligence, flat declarations about what 'Oumuamua is and isn't should be judged accordingly.

A host of theories beyond supersymmetry—the multiverse leaps to mind—are given thoughtful, respectful attention in and out of the academy despite the absence of evidence for them. That should give us pause, and not because of the absence of evidence. Rather, it should concern us because of what it reveals about the scientific enterprise itself.

What stands in our way of the fair consideration of 'Oumuamua being of extraterrestrial design isn't the evidence or the method of its collection or the reasoning behind the hypothesis. What most immediately stands in our way is a reluctance to look past the evidence and reasoning at what should follow. Sometimes the problem lies with the message, sometimes with the messenger, but when both run up against a recipient who is reluctant to listen, a problem greater than evidence and reasoning stands in the way.

• • •

There are many reasons that the search for extraterrestrial life has attracted far less attention and intellectual firepower than many of the anomalies the universe confronts us with. Certainly, the often absurd plotlines of many works of science fiction haven't helped. But neither have the prejudices of astronomers and astrophysicists — biases that cumulatively have had a chilling effect on new generations of scientists.

Today, a young theoretical astrophysicist is more likely to get a tenure-track job by pondering multiverses than by seeking evidence of extraterrestrial intelligence. This is a shame, especially because budding scientists are often at their most imaginative during the early phases of their careers. During this fertile period, they encounter a profession that implicitly and explicitly reins in their interests by stoking their fear of standing outside the mainstream of science.

An earlier generation of theoretical physicists was open to the humility of seeing their theories proven wrong by experimental data. But a new culture, one that thrives in its own theoretical sauce and exercises influence over award committees and funding agencies, is populated by advocates of popular yet unproven paradigms. When scientists double down on supersymmetry despite the Large Hadron Collider finding no evidence for it or when they insist that the multiverse must exist despite there being no data to support the theory, they are wasting precious time and money and talent. And we have not only finite funds to spend, but finite time.

The irony is that many grown-up scientists once understood this intuitively. After young children open their first checking accounts, they often fall into the trap of imagining the possible amounts of money accumulating there. As they contemplate this purchase and that purchase and all that they wish they could own, they get very excited. But after going to an ATM and learning how much money is actually in their accounts, their castles in the air come crashing down. Not only are their funds insufficient to do all that they were dreaming of, but they finally grasp the slow pace at which those funds accumulate. Usually, children will come away from this disappointment having learned to check their accounts with some frequency and balance their dreamed-of purchases against the hard evidence of confirmable data.

A scientific culture that has not learned this lesson—that does not require external verification in observable, confirmable data and that advocates for ideas deemed inherently correct due to their mathematical beauty—strikes me as a culture at risk of losing its grounding. Getting data and comparing it to our theoretical ideas provides a reality check and tells us that we are not hallucinating. What is more, it reconfirms what is central to the discipline. Physics is not a recreational activity intended to make us feel good about ourselves. Physics is a dialogue with nature, not a monologue. We are supposed to have skin in the game and make testable predictions, and this requires that scientists put themselves at risk of error.

In the age of social media, the sciences generally and astrophysics specifically need to recover their traditional humility. Doing so shouldn't be difficult. Gathering experimental data and ruling out theoretical ideas need to become greater priorities. It is reassuring to be guided by data, and it also promises more tangible, applicable rewards. Rather than spending one's entire career going down mathematical alleys that will be regarded as irrelevant by future generations of physicists, young scientists should focus on those areas of research where the value of ideas can be tested and cashed in during their lifetimes.

There is no field of research where the risk-and-reward calculus is greater than in the search for extraterrestrial life. What is more, with just eleven days' worth of accumulated data gleaned from 'Oumuamua's passage, we already have more suggestive, observable evidence than we do for all the fashionable thought bubbles that currently hold sway in the field of astrophysics.

• • •

There is value in paying attention to the intuitive leaps that children make, for they do so far more easily than many adults who carry ego-loaded baggage or intellectual prejudices. When my daughters, Lotem and Klil, learned that their father was working to send a StarChip to the vicinity of Proxima b, which sits in the habitable zone of Proxima Centauri, they grew curious, and they grew even more so when I told them the planet is expected to be tidally locked—to always show one side to the star and one side to the dark expanse of space. After a moment's thought, my younger daughter, Lotem, declared that this being the case, she would need two houses, one on the permanent night side for her to sleep in and the second on the permanent sunset strip where she could work and take her vacations.

It would be wrong to assume that Lotem's imagining of interstellar real estate was merely fanciful. Thought experiments consistent with the laws of physics are the very stuff of discovery, the means of working our way toward the solutions to the many anomalies we confront here on Earth and beyond it. In the less rigid thinking of children, we may well find the insights that propel science, and humanity, forward. And one of the worst mistakes we can make is to impose conservative presumptions on the ideas and instincts of others or laud intellectual caution for the wrong reasons.

Science is first and foremost a learning experience, one that works best by keeping us humble when we make mistakes, like children figuring out the world through their collisions with it. Just like the sharp edges of furniture, anomalies are rarely beautiful when we are first introduced to them. They confound what we think we know, stand in opposition to theories and beliefs we take as given, and resist efforts to make them align neatly with our presumptions. That is precisely when science must give priority to evidence over imagination and follow that evidence wherever it may lead.

In the late nineteenth century, for example, physicists noticed something weird about "blackbody radiation," the light emitted by heated objects. The blackbody-radiation spectrum features a single peak whose wavelength depends on temperature: the hotter the object, the shorter the wavelength of peak blackbody emission. Think of stars—small, cool dwarfs are red, warmer stars such as our Sun shine yellow, and the biggest, hottest stars are a sizzling blue. Try as they might, physicists couldn't explain or accurately model the spectral shifts at high temperatures—until 1900, when Max Planck proposed that objects absorb and emit energy in discrete units, or quanta. This revolutionary insight ushered in quantum mechanics and the era of modern physics.

No less a genius than Albert Einstein was puzzled by the weird properties of the quantum world, specifically the phenomenon of entanglement and the idea of quantum nonlocality—the mysterious ability of two particles to engage with each other no matter how far apart they are. He grappled with this unusual idea and ultimately referred to it as "spooky action at a distance." Recent experiments tell us that he was wrong to dismiss this behavior, and it turns out that the more we understand about nonlocality, the more it reveals about the very nature of reality.

Science at its core demands humility, an understanding that humanity's imagination is incapable of mapping out the full richness and diversity of nature. But the proper response to humility is wonder and, with it, a desire to open ourselves to a greater range of possibilities.

Within science, this frequently means making difficult decisions. Choices, often made outside the immediate influence of scientists, channel efforts toward certain possibilities and away from others. For example, while the number of large telescopes on Earth is constantly increasing, there are not enough of them to keep pace with the number of astronomers eager to access them. To adjudicate the competing demands for allocated time, institutions and universities have formed committees and funding agencies. They approve and prioritize submitted requests, applying the committees' expertise but also, inevitably, their biases and presumptions. I have often thought that all such decision-making bodies should automatically dedicate a significant portion of their resources—say, 20 percent—to high-risk projects. Like a financial portfolio, humanity's investments in science need diversification.

Yet many researchers stray far from this ideal, especially after they've lost their youthful enthusiasm and ascended the career ladder to tenured positions of prominence. Instead of taking advantage of their job security, they create echo chambers of students and postdocs who amplify their scientific influence and reputation. Honors should be merely the makeup on the face of academia, but they too often become an obsession. Popularity contests are outside the scope of honest scientific inquiry—the scientific truth is not dictated by the number of likes on Twitter but rather by evidence.

One of the most difficult lessons to impart to young scientists is that the search for the truth can run counter to the search for consensus. Indeed, truth and consensus must never be conflated. Sadly, it is a lesson more easily understood by a student starting out in the field. From then on, year after year, the combined pressures of peers and job-market prospects encourage the tendency to play it safe.

Astrophysics is hardly the only academic field prone to these forces, but the explicit and implicit encouraging of conservative science is both depressing and concerning, given the extent of anomalies the universe still contains. While it is not obvious to me why extraordinary claims require extraordinary evidence (evidence is evidence, no?), I do believe that extraordinary conservatism keeps us extraordinarily ignorant. Put differently, the field doesn't need more cautious detectives.

If the flame of inquiry is to continue, it is incumbent on senior scholars to not only gather to themselves promising young scholars but to cultivate an environment within which the next generations of scientists can nurture discoveries despite their inherently unpredictable nature. Budding scientists are like matches, and the context within which they do their work is like a matchbox—it does no one any good if at the moment you need them to set a new fire, they strike against the side of a worn-smooth matchbox. A long-learned professional lesson is this: if you want to nurture new discoveries, it helps to construct new matchboxes.

• • •

Scientific progress has been stifled many times over the years because the gatekeepers who established and enforced orthodoxy believed they knew all the answers ahead of time. To state the obvious, putting Galileo under house arrest did not change the fact that the Earth moves around the Sun. Centuries later, the world is unanimous in siding with Galileo. But if that is the only lesson we take from that moment in time, I worry that we will fail to learn another crucial insight. Our debts run to both Galileo and to the authorities who muzzled him. It is not enough to celebrate the first. We must also learn to guard against the second.

Surrounded by the technological comforts of the twenty-first century, scientists imagine ourselves the descendants of Galileo rather than the descendants of the men (it was entirely men) who muzzled him. But that is an error akin to a scientist cherry-picking data. Our civilization is the product of not just our scientific advances but also those moments when for any number of reasons advances were delayed or even stopped in their tracks. We stand where we stand today because of the men and women who looked through the telescope, but also because of the men and women who refused to.

Science is a work in progress, and the pursuit of scientific knowledge is never-ending. But that progress does not follow a straight path, and the obstacles encountered are sometimes of humanity's own making. Unfortunately, the humility accompanying our never-ending learning experience is, as in the case of 'Oumuamua, sometimes forgotten out of hubris, whether exercised by ecclesiastical authorities, secular authorities, or, sometimes, scientists who declare victory prematurely and assume a line of inquiry has reached its end. Examples of the last instance are myriad. A brief sampling of such moments can help us decide if we're too rapidly closing the door to every hypothesis the evidence concerning 'Oumuamua supports.

Consider that in 1894, the prominent physicist Albert Michelson argued, after surveying the great advances in physics realized during the late nineteenth century: "It seems probable that most of the grand underlying principles have been firmly established.... An eminent physicist remarked that the future truths of physical science are to be looked for in the sixth place of decimals." In contrast, over the subsequent several decades, physicists witnessed the emergence of the theories of special relativity, general relativity, and quantum mechanics, theories that revolutionized our understanding of physical reality and thus disproved Michelson's forecast.

Similarly, in August 1909, Edward Charles Pickering argued in a *Popular Science Monthly* article that telescopes had reached their optimal size, fifty to seventy inches, and there was thus little point in building instruments with larger apertures. "Much more depends on other conditions, especially those of climate, the kind of work to be done and, more than all, the man behind the gun," Pickering wrote. "The case is not unlike that of a battleship. Would a ship a thousand feet long always sink one of five hundred feet? It seems as if we had nearly reached the limit of size of telescopes, and as if we must hope for the next improvement in some other direction."

Pickering was mistaken, of course; telescopes with larger apertures collect more photons, allowing scientists to see farther out into the cosmos and deeper into the past. Pickering directed the Harvard College Observatory from 1877 to 1919, and his unfortunate words therefore carried a lot of weight, especially on the East Coast. As a result, the West Coast became the center of observational astronomy in the United States for decades to come.

It happened gradually. In December 1908 George Ellery Hale's sixty-inch telescope at Mount Wilson Observatory in California achieved first light. This was within Pickering's declared optimum range and even as Hale's telescope became more productive, Pickering and the East Coast stayed complacent. Hale, charting his own course, did not.

Hale soon built a hundred-inch telescope; it began operations at Mount Wilson in 1917 and was used by Edwin Hubble and Milton Humason shortly thereafter to determine that the universe is expanding—one of the twentieth century's seminal discoveries. That hundred-incher was the biggest optical telescope in the world until 1948, when one twice as wide came online at California's Mount Palomar Observatory. During its long career, the two-hundred-inch Palomar telescope helped astronomers discover radio galaxies and the active galactic nuclei known as quasars, which are fueled by gas falling into super-massive black holes, among many additional new sources of light.

And telescopes have just kept getting bigger, all the way up to the present day. Multiple ten-meter instruments are in operation currently, and three extremely large telescopes with apertures of twenty-four and a half meters (partnered with the Harvard College Observatory, reclaiming some of the ground that Pickering lost), thirty meters, and thirty-nine meters, respectively, are scheduled to come online in the next decade. Their large diameters will offer unprecedented angular resolution, and their large collecting areas will make them sensitive to faint, previously undetectable sources. Pickering had erred due to his arrogance. Not personal arrogance, but professional arrogance. He thought what his generation of scientists observed, understood, and determined was of interest was the peak of discovery; he didn't appreciate that the ascent of science is one false peak after another.

Unfortunately, Pickering was not unique in this particular blunder. Indeed, it is a recurrent mistake throughout the history of science. In 1925, Cecilia Payne (later Cecilia Payne-Gaposchkin) became the first Harvard student to earn a PhD in astronomy (although the degree was officially awarded by Radcliffe, as Harvard did not grant doctorates to women at the time). She concluded that the Sun's atmosphere was made mostly of hydrogen. When reviewing her dissertation, the highly respected director of the Princeton Observatory, Henry Norris Russell, argued that the Sun's composition could not be different from that of the Earth and dissuaded Cecilia from including her conclusion in the final version of her thesis. While attempting to prove her wrong in subsequent years through the analysis of new observational data, he realized that she was right.

Arrogance again retarded a field when, in the mid-1950s, Charlie Townes encountered stiff resistance as he attempted to demonstrate the feasibility of the maser (short for "microwave amplification by stimulated emission of radiation"), which would, once built, amplify radiation at a frequency particular to a given element. Two Nobel laureates, Isidor Isaac Rabi and Polykarp Kusch, visited his laboratory at Columbia University in 1954 and implored him to cease his experiments on ammonia, insisting that the device would never work. Luckily, Townes persevered, and masers became the timekeeping devices in atomic clocks and were widely used in radio telescopes and deep-space spacecraft communication. In collaboration with a number of scientists, Townes did pioneering work with masers that led directly to the development of lasers.

Here's an even more recent example. I once asked a prominent astronomer who studies objects in the Kuiper Belt—the ring of icy bodies beyond Neptune's orbit—if he was looking for brightness changes way out there that might indicate artificial light. He scoffed at the very suggestion: "Why? There is nothing to look for."

The establishment initially regarded Kuiper Belt objects (KBOs) as imaginary constructs. Pluto was the exception, of course; the largest KBO, it was discovered by Clyde Tombaugh in 1930 and thought to be a planet. But more than half a century later, UCLA astronomer David Jewitt couldn't get telescope time or funding to hunt for KBOs, so he piggybacked his search with other projects. In 1992, he and Jane Luu finally discovered the first non-Pluto KBO using the eighty-eight-inch telescope at the top of Hawaii's Mauna Kea.

In each instance, a leap forward was forestalled, and it was not because of a lack of available technology or an absence of imaginative curiosity or the unavailability of testable data. The delay stemmed from the arrogance, often well intended, of influential gatekeepers. And as much as we now celebrate the wonders of ever larger, grander telescopes and the world of possibilities they open up, what might have followed had scientists made these discoveries years, generations earlier?

• • •

Many scientists see themselves as a breed apart, members of an elite intelligentsia. Consciously or subconsciously, they want to separate themselves from the rabble. Such thinking motivates, at least in part, an argument made by many scientists I know: that scientists should communicate with the public only after they figure something out to a certainty. If laypeople knew the messy reality of science—that it's full of starts and stops and dead ends—they'd brand every result as preliminary or questionable, the reasoning goes. Every important scientific consensus—such as the effect of human beings on Earth's climate and the potentially disastrous consequences for us and all other life on the planet—could, some scientists fear, be summarily dismissed. This withholding strategy has the added benefit of making scientists look smarter than we actually are, and, adding to its appeal, it limits outside criticism.

But this approach is wrong. Keeping the public informed is our duty, and not just because so much scientific research is taxpayer-funded. A public that is deeply informed, engaged, and enthusiastic about scientific advances is a public that directs not just its financial support but the interest and efforts of its children, its brightest minds, toward the most confounding challenges. In that spirit, being more open as to what we know and what we do not will increase scientists' credibility over the long haul. Shutting the public out until the very end can also lead to mistrust. After all, the anomalies we confront are not for scientists alone. They confront all humanity, and when there are breakthroughs, much like medical advances, it is to the benefit of everyone. We should show the world our work in progress, especially when it is full of uncertainties and buffeted by competing interpretations due to lack of conclusive evidence; we should let everyone see how surprised we often are at what we find.

Also, the general opprobrium with which the academy has met undergraduate interest in SETI has had a chilling effect on graduate-student interest. By one estimate, only eight scholars worldwide have completed doctorates in transparently SETI subjects. But that may be changing a little. As I write, seven graduate students are currently on track to receive PhDs in SETI-related subjects. What sorts of questions and, consequently, what sorts of experiments in pursuit of what kinds of data ought we to encourage in the next generation of astronomers? Here, again, 'Oumuamua nudges us, if we care to pay attention. The traffic of technological equipment through interstellar space might be startling but it's not noticeable until we develop instruments that are sensitive enough to detect it.

Indeed, I have sometimes—with a tip of the hat to Yuri Milner—described the search for extraterrestrial life as the ultimate venture-capital investment of scientific research. Any search method is, like investments, risky. With SETI, we have few clues regarding the properties of the needle we are searching for in the haystack of the universe, but if any needle is found, the reward will be tremendous. The return on such an investment would greatly eclipse other, narrower scientific interests. Just knowing that we are not alone would transform humanity itself, to say nothing of the knowledge we might gain from such a discovery.

I know it can be hard, especially for young scientists, to advocate for ideas deemed outré by the establishment. At this point in my life I have considerable career stability as well as an inherent disinclination—one that traces back to at least my first day of first grade—to seek the approval of others. Even so, I might not have been ready to advance the 'Oumuamua-is-an-alien-lightsail hypothesis—or explore the possibilities it contained—had not the fragility of life and the window of time each of us individually has to advance the common good been brought keenly to my attention. For this, my scientific work on the universe was partly, although not entirely, to blame.

8

Vastness

WHEN YOU ARE IN THE MIDDLE OF READING A Sherlock Holmes story, it is easy to forget the vantage point of Holmes. For him, any individual case is, well, just one among many. And Holmes's observation "Eliminate all other factors, and the one which remains must be the truth" applies to his habits of deduction whether he says it in *The Sign of Four,* "The Adventure of the Beryl Coronet," "The Adventure of the Priory School," or "The Adventure of the Blanched Soldier."

In this way, productive astrophysicists are not unlike fictional detectives—while not all anomalies are the same, the process of trying to unravel them is.

"Eliminate all other factors," orders Holmes. And there is, as it happens, another factor that bears on the question of 'Oumuamua's origins and purpose. It has to do not with 'Oumuamua itself but with the universe that it is touring—a universe that is older and vaster than anything else we know. Its very ancientness and vastness, indeed, may hold the key to unlocking another one of 'Oumuamua's mysteries.

• • •

During a family vacation to Cradle Mountain in the highlands of central Tasmania a decade before Oumuamua was discovered, I stepped outside after dinner and looked up. Because we were so far removed from the hubs of civilization, there was none of the usual light pollution that spoils the view from so many of the world's backyards. I stared into a clear night sky.

It was overwhelming. Arrayed overhead were the countless stars of our galaxy, the Milky Way, stretching across the sky. Off to the side, I could see the Large Magellanic Clouds and our nearest galactic neighbor, Andromeda, a twinkling, iridescent patch that appears roughly the size of the moon. Some of my joy in the sight was in appreciating the fact that it was not timeless. It is anyone's guess if humanity will be around to witness it, but it is a certainty that what we look up and see tonight is no more eternal than we are.

At the time, I was especially sensitive to the universe's impermanence. Just a few years earlier, I had had the original idea of simulating the future collision between the Milky Way and Andromeda. I was particularly fascinated with our distant cosmic future, following on earlier papers where I showed that the accelerated expansion of the universe will leave our home galaxy in a void of empty space. Once the universe ages by a factor of ten, all distant galaxies will be pulled away from us faster than the speed of light, and humanity will be able to observe only the stars in our own galaxy. What would this galaxy look like? Aside from transforming the appearance of the night sky, the gigantic collision with Andromeda could kick the Sun to the outskirts of the merged galaxy and establish our new cosmic neighborhood for ten trillion years to come, until the light from all stars, including the faintest and most abundant dwarf stars, like Proxima Centauri, is extinguished. I convinced my postdoc T. J. Cox to simulate this future collision, and we reported in 2008 that within a few billion years, long before the Sun dies, our night sky will change and the stars from the two sister galaxies will mix to make a new, football-shaped galaxy, which we named Milkomeda.

It was remarkable to recognize, that night in Tasmania, the objects of my studies. The Milky Way and Andromeda galaxies were splashed across the sky in a brilliant cascade of light. Perhaps as a result of seeing them so clearly, I felt my place among them more keenly than I usually did. This is a pleasure of astronomy. By contrast, particle physicists do not have the privilege of seeing the Higgs boson with the unaided eye.

But my thoughts that evening were not occupied only with the transformation of our galaxy in the distant future. Uppermost in my mind was the question of how the first generation of stars and galaxies lit up in the very beginning of our universe, the scientific details of the story of the universe's genesis.

The cosmic dawn was my first fascination as an astrophysicist. My interest began during my time at Princeton and sharpened with the passage of years. Eventually, my investigation of this mystery would influence my pursuit of another, shaping my thinking not just about the history of our universe but also about any civilizations with which we may have shared it.

When you gaze up on a clear night, as I did all those years ago in Tasmania, the numerous sun-like stars of the Milky Way look like the lights in the main cabin of a giant spaceship streaming through the universe with passengers next to some of these lights. What can we learn about these passengers based on our brief encounter with 'Oumuamua? What, for that matter, can we learn about ourselves?

• • •

We date the birth of the universe, the Big Bang, to some 13.8 billion years ago. Fascinating, revelatory work has been done that has produced theory, data, and confirmed predictions concerning the universe's earliest origins, including the common agreement that after the first hundred million years, everything was cloaked in darkness. Until, that is, the first star was born.

How did the earliest stars come into existence? After my arrival at Harvard in 1993, with Zoltan Haiman, my graduate student, and Anne Thoul, my postdoc, I worked out a theory to explain their formation.

Following the Big Bang, matter was spread more or less evenly across the rapidly expanding universe. That it was spread *almost* uniformly is crucial, for in some places, we theorized, the cosmos started out slightly denser than average. "Slightly" in the sense of one one-hundred-thousandth more dense. But those slight perturbations were sufficient. That was enough for gravity to start to pull matter into these increasingly dense regions and for clouds of gas, composed mostly of hydrogen atoms, to begin to assemble.

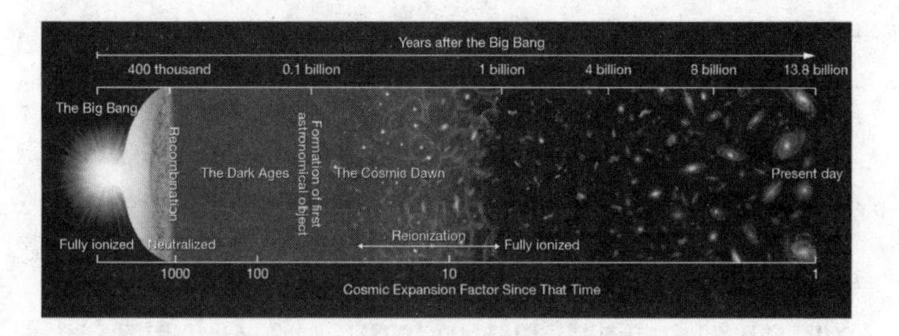

Timeline of the history of the universe. The solar system formed relatively late, only 4.6 billion years ago. Modern technology appeared on Earth only over the past century, 0.0000001 billion years ago. Many civilizations could have appeared and disappeared before we developed our modern telescope technologies to detect them. Image by Mapping Specialists, Ltd.

Working with pencil and paper, my research team modeled the idea until we reached the point where only more sophisticated computer hardware could advance it further. Volker Bromm, a graduate student at Yale at the time, took on that task, and over the past two decades he and other theorists have established that, indeed, the process we outlined for the birth of stars could give rise to the early galaxies. Models and theories are invaluable, but data that proves both remains essential. I wanted to see the gas clouds our theory predicted, which meant trying to find evidence that was about thirteen billion years old.

When astrophysicist-detectives confront the challenges of scale that the universe presents, they can become overwhelmed. They do, however, have one asset without parallel in any other academic discipline. They have the ability to look back in time.

Because light travels at a finite speed, the farther out we look, the farther back in time we see. And since the universe had similar conditions everywhere, by peering deep into space, we can view our own past.

The deeper into space we look, moreover, the older the objects we uncover. To look at a star four light-years away, such as Proxima Centauri, is to look at a star as it was four years ago. But if we focus our telescopes on a galaxy that was thirteen billion light-years away when its light was emitted, we are glimpsing the universe as it was thirteen billion years ago. To peer that far back into the "dark ages" of the universe, the moment when the clouds of gas from which the first stars arose gathered, is a monumental scientific challenge. It also forces us to contemplate the incomprehensibly vast timescales of the universe. Humans today live, on average, nearly seventy-three years. To have seen the first lights in the universe come on thirteen billion years ago, we would have had to live almost one hundred and eighty million lifetimes—an idea that is especially absurd considering that the Earth is only about 4.5 billion years old and that we believe the planet has supported life for only 3.8 billion of those years.

Looking into the universe, astrophysicists also come face to face with the physical immensity of the cosmos. We can see light that was emitted earlier in cosmic history. The universe resembles an archaeological dig centered on us. The deeper we look, the more ancient are the layers we uncover. This exhibit of cosmic history continues all the way back to the edge of the visible sphere around us, located at the Big Bang, 13.8 billion years of light travel time. It takes light originating beyond this edge more than the age of the universe to reach us, and so more distant regions are not visible to us.

It is very presumptuous for us to assume that we are the only intelligence in this vast cosmos. Even though life as we know it and life as we do not know it may exist on numerous other planets, it is most likely that we will encounter relics of extraterrestrial technologies before establishing contact with any living civilization. This must be kept in mind as we contemplate explanations for the mysterious properties of interstellar objects like 'Oumuamua.

• • •

My research on the cosmic dawn contributed to the creation of a new field of study, what has come to be called "twenty-one-centimeter cosmology." It is a branch of radio astronomy that maps the universe in three dimensions using radiation from hydrogen atoms that started out with a wavelength of twenty-one centimeters that was subsequently stretched by cosmic expansion.

You may recall that this is the same meter-wave radio spectrum that humans fill with the noise of their televisions, radios, cell phones, and computers—an insight that inspired Matias Zaldarriaga and me to wonder if other advanced civilizations would likewise emit such noise. But my initial interest in twenty-one-centimeter emissions was as a means to stare back to a time well before any civilization was possible. At this phase of my career, I wasn't hunting aliens; I was hunting hydrogen.

After the Big Bang, hydrogen was the most abundant element in the universe by a wide margin; the early universe was about 92 percent hydrogen atoms and 8 percent helium atoms. But at this point, the hydrogen in the universe wasn't emitting any radio signals that we can detect today. That is because in the immediate fiery aftermath of the Big Bang, the vast majority of the ordinary matter of the universe, hydrogen, was ionized.

Neutral hydrogen atoms consist of a single proton and a single electron. But at high temperatures and with intense ultraviolet radiation, they are broken (ionized); the hydrogen atom sheds its electron and exists as a single positively charged proton. This changes hydrogen's behavior—or, more accurately, the type of radio signal it emits. The electron bound to a neutral hydrogen atom can transition between energy states, higher and lower, and in doing so emit a photon, or light particle, in the form of a radio wave twenty-one centimeters long. But ionized hydrogen cannot.

About three hundred and eighty thousand years after the Big Bang, the universe cooled down enough for electrons and protons to combine and form neutral hydrogen atoms, and we can start to seek the element's telltale signature, twenty-one-centimeter radio waves. For hundreds of millions of years, hydrogen atoms remained neutral, passing between their higher and lower energy states and emitting waves right up until stars and then galaxies began to form—and the hydrogen in the universe was ionized all over again.

Stars emit more than visible light; they also emit ultraviolet radiation, which can split hydrogen atoms into their component parts, electrons and protons. When the first stars turned on, they re-ionized the universe's neutral hydrogen atoms. This was less a moment than an epoch, a long period when ultraviolet light from the earliest stars and black holes split the universe's dark fog of neutral hydrogen into protons and electrons. But the changing chemistry of the universe gave astrophysicists data to search for

—namely, the absence of twenty-one-centimeter emissions. Ionized hydrogen atoms do not emit these radio signals, but neutral hydrogen atoms do.

Thus, the moment when the twenty-one-centimeter emission signal disappears is the moment the stars were born. Like the famous Sherlock Holmes story that turned on the dog that did not bark, this scientific mystery became the case of the hydrogen that no longer produced its twenty-one-centimeter emission.

As I write, the search is under way for data that will help us pinpoint when, exactly, the stars began to shine. In South Africa, a multi-antenna array called the Hydrogen Epoch of Reionization Array (HERA) is currently measuring twenty-one-centimeter emissions from the early universe. The Hubble Space Telescope recently identified a galaxy that flipped its lights on just three hundred and eighty million years after the Big Bang. And the James Webb Space Telescope—whose first science advisory group I served on, decades ago—is expected to launch in 2021 and be able to find galaxies at even earlier times. Under development are the twenty-four-and-a-half-meter Giant Magellan Telescope, the Thirty Meter Telescope, and the European Extremely Large Telescope, which has an aperture diameter of thirty-nine meters.

Our contact with the data from these efforts has only begun and with it the winnowing of explanations for how the stars came to shine. And the answer, when we discover it, will bear immediate relevance to the question of whether intelligence other than our own is out there in the expanse of space. If ʻOumuamua is alien technology, then it is a near certainty that its designers also looked into the dim past of our common universe and likewise teased out meaning from ionized and neutral hydrogen. To be curious enough to explore space in the neighborhood of one's own solar system or out among the stars is by definition to be curious about the universe—what its properties are, what explains its past, and what predicts its future. It is not just that our own curiosity and behavior is our best guide to the curiosity and behavior of extraterrestrial life. It is also true that the insights of science will provide us with the common language we need to make sense of extraterrestrial intelligence, perhaps even communicate with it. Science also provides us with a means of making sense of what we discover, however fleeting, however partially. For if we can build it, the odds are great that another intelligence, if it is out there, has done the same.

9

Filters

IF THE LIGHTSAIL HYPOTHESIS IS TRUE, THERE ARE two possible explanations. One is that ʻOumuamua's makers intentionally targeted our inner solar system; the other is that Oumuamua is a piece of space junk that happened upon us (or we upon it). Either of these interpretations could be accurate regardless of whether the civilization that created ʻOumuamua still exists today. But given what we know about the universe and about civilizations, we can make some inferences about which interpretation is likely correct and what implications it holds for us and for whoever (or whatever) created ʻOumuamua.

The space-junk idea is similar to the asteroid/comet hypothesis in an important way: it implies that ʻOumuamua is part of an incredibly huge population of similar objects. Every star in the Milky Way would have to send, on average, a quadrillion of these things into interstellar space in order to make it conceivable that one would happen to zip past our telescopes just as we trained them on the sky. That translates to one launch every five minutes from every planetary system in the galaxy, and it assumes that all civilizations live as long as the thirteen-billion-year-old Milky Way, which is certainly not the case.

The idea that civilizations could manufacture their way to such a density of objects, critics argue, seems even more unreasonable than all the conjectures concerning planet formation and the release of material from outer clouds to produce a sufficient population of rocks. To fill the universe with space junk at such a density, a great many civilizations would have to spend a great deal of time ejecting a great deal of material. Of course, the moment we posit an intelligence behind some materials' construction, we also do away with the need for a random distribution of materials. After all, we did not send our five interstellar rockets off on random trajectories. Scientists decided to send them toward certain stars, and we can anticipate that another intelligence would do the same.

We should also avoid the trap of imagining interstellar spacecraft as rare and precious, as our paltry five interstellar probes could suggest. Given the rarity with which humanity has sent material out into interstellar space, the hypothetical abundance I have postulated would seem unreasonable indeed.

This scenario seems a little less unreasonable if we think of this possibility against the potential projected rate of ejecting StarChips using the Starshot Initiative that my colleagues and I proposed to Yuri Milner. We estimate that once the investments have been made to build a suitably powerful laser and launch it into space, the relative costs of sending many thousands, indeed millions, of StarChips into interstellar space drops exponentially.

But the abundance of interstellar spacecraft in the scenario I've just described will seem most reasonable, perhaps, if we return to our plastic bottle.

• • •

Right now, the United States Space Surveillance Network tracks more than thirteen thousand man-made objects orbiting Earth. These include everything from the International Space Station to defunct satellites, from orbital telescopes like Hubble to discarded rocket stages and even nuts and bolts left behind by astronauts. It also includes roughly twenty-five hundred satellites it has taken us fifty years to put into space.

Indeed, during that brief window of time, our efforts to send material up into our planet's orbital plane have been sufficient to make space junk a looming problem. For example, in 2009, two satellites, Russia's inactive Cosmos 2251 and America's active Iridium 33, collided at approximately 22,300 miles per hour above Siberia. The result was an instantaneous cloud of debris, which increased the risk of more collisions. This was the first known collision between satellites and it underscores the danger of a rising amount of junk orbiting Earth.

The threat of collision has been steadily increasing for years, in part due to ever more nations viewing space as a new frontier for conflict. Over a decade ago, China demonstrated the success of its anti-satellite missile technology by destroying its own Fengyun-1C weather satellite. India accomplished a similar feat in 2019, creating another four hundred pieces, give or take, of space debris. A consequence? The risk of impact to the International Space Station was estimated to have gone up 44 percent over ten days. No wonder the station is designed to maneuver out of the way of danger —assuming it has enough warning.

What humans do helps us predict what other civilizations are likely to do. We continue to be our own best data set for imagining the behavior of other civilizations and the consequences of that behavior. With that in mind, consider that a computer simulation looking out two hundred years predicts we will act in ways that will multiply the amount of space-junk objects that are larger than about eight inches by a factor of 1.5. And smaller junk will increase even more. The simulation predicts that the number of objects less than four inches will increase by a factor of somewhere between 13 and 20.

This junking-up of space is, sadly, in keeping with humanity's treatment of its terrestrial habitat. In 2018, the World Bank issued a report entitled "What a Waste 2.0" in which it estimated that the world generated 2.01 billion tons of solid waste a year. The World Bank also projected that by 2050 that number could go as high as 3.4 billion tons. In 2017, the U.S. Environmental Protection Agency estimated that the average American generates 4.51 pounds of solid waste a *day*, and the United States is far from the greatest producer. While the United States and China produce the most greenhouse-gas emissions, it is the lower-income countries that produce the most solid waste due to their inability, driven by economics, to properly dispose of it.

Of course, from the vantage point of Earth itself, the *origins* of the world's solid waste doesn't matter. Much of it ends up in the oceans regardless.

One of the fastest-growing areas of waste is what is termed e-waste—discarded laptops, mobile phones, and home appliances that have been displaced by newer models. In 2017, the United Nations' Global E-Waste Monitor estimated that in the prior year, the world had generated 44.7 million metric tons of electronic waste. And it estimated that by 2021, that would rise to 52.2 million metric tons.

Here our own civilization's behavior offers, once again, another piece of evidentiary data we can consider when we wonder about 'Oumuamua's origins. If we suppose 'Oumuamua was not a functioning probe or an inert buoy but rather another civilization's defunct or even discarded technology, that suggests that another civilization acted in ways we can immediately identify with—that they were, like us, profligate in their production of materials, technological and otherwise, and that, like us, they were comfortable abandoning them when obsolete. Just because we have not yet reached the maturity to discard materials into interstellar space should not blind us to the possibility that our interstellar neighbors might have or, more likely, did.

Trash, both in its solid-waste form and its greenhouse-gas-emission form, is a useful analogy for a different reason: it suggests an answer to the question of how ʻOumuamua might have ended up roaming the universe as space junk. Because one of the insights granted by pioneering physicists in this field—men such as Frank Drake, whose famous equation quantifies our chances of detecting a light signal from an advanced civilization in space—is that most of the technological civilizations that ever existed might now be dead.

• • •

Enrico Fermi was one of the giants of twentieth-century physics. Among his accomplishments is the development of the first nuclear reactor and, as he was instrumental in the Manhattan Project and the production of the first nuclear bomb, he can claim some credit for the prompt ending of hostilities with Japan at the conclusion of World War II.

Toward the end of his storied career, during a lunch with his colleagues, Fermi raised a simple, provocative question: How do we explain the paradox that, given the vastness of the universe, the probability of extraterrestrial life seems high, yet there is no certain evidence for anything but terrestrial life? If life is common in the universe, he asked, "Where is everybody?"

Over the years, many answers have been formulated. One is especially arresting and especially pertinent to the unfolding mystery of ʻOumuamua and its implications for us.

In 1998, the economist Robin Hanson published an essay titled "The Great Filter—Are We Almost Past It?" Perhaps the answer to Fermi's paradox was, Hanson argued, that throughout the universe a civilization's own technological advancement overwhelmingly predicts its destruction. The very moment when a civilization reaches our stage of technological advancement—the window where it can signal its existence to the rest of the universe and begin to send craft to other stars—is also the moment when its technological maturity becomes sufficient for its own destruction, whether through climate change or nuclear, biological, or chemical wars.

Hanson's thought exercise has sufficient plausibility that humanity would do well to consider the question in his article's title: Is human civilization nearing its own filter?

It would be no small irony if Fermi is the solution to his own paradox, for, with Fermi's help, we developed nuclear weapons seven decades ago. But even without nuclear weapons, we are moving to destroy ourselves by permanently changing the climate. The rise of antibiotic resistance, due to many factors but certainly including the largely indiscriminate use of antibiotics in industrial agriculture and livestock, also poses a threat. So do pandemics, accelerated and exacerbated by our industrialized assault on our planet's ecosystem.

It is quite conceivable that if we are not careful, our civilization's next few centuries will be its last. If that's the case, the emissions we've been sending out into the universe from our radios and televisions—that outward expanding bubble of noise humankind started to generate only a century ago—and the five interstellar craft we have already launched could well end up the equivalent of dinosaur bones here on Earth, evidence of something once formidable and extraordinary that is now only material for other civilizations' archaeologists.

We need not look far to appreciate how the great filter might work. The small filter of our own mortality and the context of recent history provide useful data.

My father's family had its roots in Germany for seven centuries. My grandfather Albert Loeb fought valiantly in World War I and survived the Battle of Verdun in 1916. That one battle, the longest of the war, is estimated to have killed 143,000 German soldiers out of a total death toll of 305,000. The dead and wounded military personnel from the entire war ranges between fifteen and nineteen million; add civilian casualties and that number rises to some forty million.

My grandfather distinguished himself in the cavalry during that conflict and was awarded a medal that a decade or so later meant little. At a town gathering held in 1933 in the district of Netze-Waldeck, where my grandfather's family lived, a member of the Nazi Party loudly argued that the country's Jews were using up Germany's resources. My grandfather stood up and confronted the man: "How dare you say these words when you personally dodged the draft in the Great War as a Communist while I was on the German front?" The speaker replied: "We all know about your patriotic contributions, Mr. Loeb; I was talking about the other Jews." But the rising tide of vicious anti-Semitism in Germany, and indeed in much of Europe, was discernible.

It was after that public exchange that my grandfather decided to leave Germany. He threw away the medal and, in 1936, emigrated to what was then British-controlled Palestine and is today Israel. Other branches of the family stayed in Germany, believing that they could wait and see what happened, holding to the belief that they would still be allowed to leave on the last trains out of Germany. Unfortunately, by then, those trains led elsewhere and all sixty-five of our family members were killed in the Holocaust.

I still keep Albert's pocket watch from a century ago in memory of his courage and integrity. It carries the same initials as mine, which is a reminder of sorts as well. The chain of causation that brought us here is tenuous indeed.

• • •

The mystery of 'Oumuamua began shortly after my father died, in January 2017, and unfolded as my mother's health declined. She was diagnosed with cancer in the summer of 2018 and passed away in January 2019.

My father, David, was laid to rest in the same red soil in which he planted trees all his life, in the vicinity of those plantations that he watered routinely, near the house that he built with his rugged hands and that I grew up in, surrounded by the people he loved and who loved him in return, under the blue sky that I study as an astronomer. My mother, Sara, who put me on the road to thinking as a philosopher, with whom I spoke daily throughout my adulthood, and who especially gifted me with the life of the mind, was buried beside him two years later.

In astronomy, we realize that matter takes new forms over time. The matter we are made of was produced in the hearts of massive stars that exploded. It assembled to make the Earth that nourishes plants that feed our bodies. What are we, then, if not just fleeting shapes taken by a few specks of material for a brief moment in cosmic history on the surface of one planet out of so many? We are insignificant, not just because the cosmos is so vast, but also because we ourselves are so tiny. Each of us is merely a transient structure that comes and goes, recorded in the minds of other transient structures. And that is all.

The deaths of my parents brought home to me this and other fundamental truths about life. We are here for a short time and consequently we had better not fake our actions. Let us stay honest, authentic, and ambitious. Let our limitations, very much including the limited time we are each given, encourage humility. And let us allow the small filter, which represents the extent of our own lives, give us an approachable and sobering context for Hanson's great filter, which represents the end of our civilization. With insufficient care, diligence, and applied intelligence, humans have proven themselves all too comfortable with ending the lives of their fellow humans.

Of all the lessons we can learn from ʻOumuamua, the most essential might be that we cannot allow the smaller filters of war and environmental degradation to grow into a great filter. We must exercise greater care, diligence, and applied intelligence in the preserving of our civilization. Only in this way can we save ourselves.

During my years of military service, there was a phrase we were taught during infantry training: *to lay your body on the barbed wire.* Sometimes, faced with extraordinary circumstances, a soldier must deliberately lie down on barbed wire to allow his fellow soldiers to use his body to safely cross over. I am not so grandiose as to imagine my experience is the equivalent of such a soldier's sacrifice. But, mindful of the specter of the great filter and ever mindful of the shadows of those who have come before and in their turn have sacrificed to advance humanity's common cause, I find the image inspiring.

Of this I am certain: The tenuous threads connecting humanity's Earth-bound civilization as it exists today, and the promise of humanity's possible interstellar civilization as it might exist tomorrow, will not be upheld by exercising conservative caution. In the words of Rebbe Nachman of Breslov, "The whole world is nothing but a very narrow bridge, and the key is not to be fearful at all."

• • •

On September 1, 1939, three years after my prescient grandfather left Nazi Germany, Germany invaded Poland, and much of our planet found itself at war. It would be another eight months before Winston Churchill assumed his role as the wartime prime minister of the United Kingdom. In the interim, Churchill relentlessly warned his country and the world of the threat posed by Adolf Hitler and a militaristic Germany.

Churchill also continued one of his cherished pastimes: writing. That decade, he penned, among other things, a four-volume biography of the first Duke of Marlborough and numerous opinion pieces and articles for newspapers and magazines. A subject of special interest was science (Churchill became the first British prime minister to appoint a civilian science adviser to the government), and his popular-science essays touched on everything from evolution to fusion power to aliens.

In 1939, as the world was collapsing around him, Churchill penned an article titled "Are We Alone in Space?" He never published it; the confluence of events that would bring him to the apex of his political influence would also sweep this essay aside and bury it for decades. A war was fought and won before Churchill, once more out of political fashion in the United Kingdom, revisited his article. In the 1950s, he gave it the more accurate title "Are We Alone in the Universe?" But it still sat unpublished at the time of Churchill's death, and it entered the U.S. National Churchill Museum archives unknown and uncommented upon until its discovery in 2016.

It is a shame that Churchill's unusual essay was never published, for it contains ideas that were well ahead of their time and a sense of perspective that was sorely needed then, as it is now. Churchill evinced the modesty of a generalist as he wondered how unique the Sun and our planetary system might be, writing, "I am not sufficiently conceited to think that my sun is the only one with a family of planets." He was also astute. Decades before the discovery of exoplanets, Churchill concluded it was reasonable to believe that a large number of planets existed "at the proper distance from their parent sun to maintain a suitable temperature," had both water and atmosphere, and so could potentially support life. Indeed, given the expanse of space and the number of suns in it, he wrote, "The odds are enormous that there must be immense numbers which possess planets whose circumstances would not render life impossible." And though skeptical of interstellar travel, Churchill allowed that "in the not very distant future, it may be possible to travel to the moon, or even to Venus or Mars."

The somber note in the article isn't reserved for the possibility of extraterrestrial life in the universe or even of humanity's ability to reach other planets but rather for humanity itself. "I, for one, am not so immensely impressed by the success we are making of our civilization here," Churchill wrote, "that I am prepared to think we are the only spot in this immense universe which contains living, thinking creatures, or that we are the highest type of mental and physical development which has ever appeared in the vast compass of space and time."

When I first heard about Churchill's essay, a few years ago, I couldn't help but indulge in a thought experiment. The planet-spanning war that erupted shortly after Churchill wrote his essay is estimated to have cost $1.3 trillion—the equivalent of about $18 trillion in today's currency. No reliable records exist to accurately estimate the number of people who died due to the war, and scholars bicker over which deaths can be clearly attributed to the war itself, but the range falls between forty million and one hundred million human beings.

What if in the 1940s humanity had instead spent that $1.3 trillion—not to mention the skill, expertise, muscle, and minds of between forty and one hundred million people—on the exploration of the universe? What if the era's collective genius hadn't turned itself to destruction, efforts that reached their pinnacle in the development of nuclear weaponry, but had turned instead toward sending terrestrial life into the solar system and then out into the expanse beyond? What if human civilization, out of humility and the application of the scientific method, had concluded that its own existence made the existence of other civilizations in the universe probable? What if, in 1939 and over the course of the next decade, humanity had oriented itself toward space exploration and the discovery of extraterrestrial life rather than the vast extermination of life here on Earth?

If there is a multiverse and if such a version of human civilization exists in it, I predict that at a minimum, it managed to photograph ʻOumuamua, maybe even capture it for thorough examination. Perhaps those humans were not even surprised by what they found, for on their version of Earth, the Breakthrough Initiative would have started decades earlier and as a consequence, they would already be in receipt of the information that laser-powered lightsail ships had captured during their passage near Proxima Centauri. They would be well on their way, surely, to contemplating a solution to ensuring life's continuation after the inevitable death of our Sun. I suspect, too, that their beaches would be less littered with trash.

I'm sure there is one similarity, at least, between that Earth and this one. I will bet that their historians refer to the pivotal generation — the one that, in the 1940s, set it all in motion — as their greatest.

Alas, we reside on *this* Earth, collectively tasked with the preservation of our human civilization. Among all the thought experiments that the multiverse theorists give us, the most productive, I believe, is this one: What will we do as residents of the one universe immediately arrayed before us?

As I write, I consider the tree visible from my living-room window. Are we a civilization that will bind the damaged limb, allowing it to mend and grow? Or are we a civilization that either ignores it or shears it off, forever ending that branch of possibility?

Whichever choice we make, we bet with our children's children's lives. If, when confronted with ʻOumuamua's exotic features, the only hypotheses we can contemplate are the natural but statistically remote — if we cannot, as Sherlock Holmes might, entertain the simplest remaining explanation to the collected data — we may do worse than simply delay our civilization's next leap forward. We may walk into the abyss, one civilization among many and perhaps one not even advanced enough to have left, as a calling card, buoys throughout the universe.

10

Astro-Archaeology

IF WE CONCLUDE THAT CIVILIZATIONS WINK IN AND out of existence, perhaps serially, over the long, long history of the universe, it would be a grim warning for our own civilization.

It would also be an opportunity.

As scientists and as a species, we could tailor our detective work and search for the relics of dead civilizations. Even an oblique discovery of such evidence could teach us an important lesson—namely, that we need to get our act together if we're to avoid a similar fate.

As I've mentioned, this could prove to be 'Oumuamua's profound message in a bottle, one that we're stubbornly refusing to read. Grappling with it fully, I believe, will require us to stop thinking of astronomy simply as the study of stuff in space and start treating it as an investigative and interdisciplinary enterprise.

We are greatly in need of a new branch of astronomy, what I have termed *space archaeology*. Similar to archaeologists who dig into the ground to learn about, say, Mayan society, astronomers must start searching for technological civilizations by digging into space.

It is enthralling to imagine what these astro-archaeologists might find, but that's not even the most compelling reason to take this research seriously. It could well hold insights that nudge us in new scientific and cultural directions — and perhaps make our civilization one of the rare ones to outrun the great filter.

• • •

Recall that one of the biggest limitations of the Drake equation — the formula designed to help anchor discussions about intelligent extraterrestrial life — was its myopic focus on communications signals, which are only one of the various detectable traces that other civilizations might leave behind. Frank Drake defined the first variable of his equation, *N*, as the number of species in our galaxy that possess the technology necessary for interstellar communication; he defined the equation's final variable, *L*, as the duration of time such species are able to produce detectable signals. In short, his equation was bracketed by the assumption that intentional efforts to communicate were the only way in which extraterrestrial civilizations would be detectable.

However, there are lots of ways in which alien civilizations might unintentionally broadcast their existence, and as we discover new technologies, the number of new avenues available to search for this evidence is increasing. How ought we to redefine the scope of our search? To put it another way: What should we be looking for? And where should we be looking?

The first of these questions is, I believe, relatively easy to answer. We know that life-forms of all sorts are identifiable by what we call biosignatures — for example, algae blooms and polluted atmospheres that living things leave in their environment. So in addition to seeking traces of technologically advanced extraterrestrial life, we can search for evidence of less advanced alien life, such as microbes, whether living or long dead.

Thus the first question leads to another, more granular question: What kind of life should we be looking for, advanced or primitive? In a paper that I wrote with my postdoc Manasvi Lingam, we estimated the likelihood of finding primitive versus intelligent extraterrestrial life using only state-of-the-art telescopes (at the time, that included the James Webb Space Telescope, the successor to the Hubble Space Telescope). At heart, this was an effort to determine what part of astro-archaeologists' efforts should be spent on the search for biosignatures and what part on the search for technosignatures. The exercise helped sharpen my thinking about the answer to the question I posed above: What should we be looking for?

This project required us to work through a number of highly uncertain variables, including several that demanded best-guess answers. For example, we had to determine how much rarer intelligent life was than microbial life, how much farther away we could detect a technosignature versus a biosignature, and how long both types of signatures would be discoverable. Our selection of variables also reflected our concern over the great filter, although we made an optimistic guess about how long the sort of extraterrestrial technological intelligence we were seeking would survive—we set it at a millennium.

Optimism is a precondition for scientific work, I've found, but in this case, optimism also factored into our actual calculations. In more ways than one, the more pessimistic you are, the lower the prospect of finding intelligent life. Consider, too, that in the scenario I have just described, we needed to guess both the duration of time in which intelligence might be discoverable as well as another, related variable: the period over which *our* intelligence would be around to seek it.

With that said, it must be admitted that discovering primitive, or microbial, life would not be the same as discovering intelligent extraterrestrial life. Either would fundamentally alter humanity's view of itself, but evidence of technological intelligence would have the greater impact. To learn that other, perhaps even more advanced intelligent civilizations exist or preceded us would force us to adopt a humbler attitude toward the universe and our accomplishments.

Ultimately, we concluded that the likelihood of detecting intelligent life was approximately two orders of magnitude smaller than the likelihood of detecting primitive life. Yet we also concluded that both searches should be undertaken concurrently, if with significantly more funding dedicated to primitive life, given that we expected it to be more plentiful. Additionally, the existence of intelligent life would greatly raise the prospects of our finding microbial life too.

So what should we be looking for? In a word: *life*. We should just be prepared to find one type of it sooner than the other.

Where, then, should we be looking? Answering this question is trickier and more complicated—but ultimately, perhaps, more comfortably familiar. Because it requires us to start with terrestrial abiogenesis: the origin of life on our planet.

• • •

The field of origins-of-life research is nascent. While we know a great deal about one aspect, terrestrial abiogenesis, our knowledge is an island in a vast ocean of ignorance. Yet there are reasons to be cautiously optimistic about where it is headed.

As I write these words, we are far closer to understanding how the first cells, the building blocks of life, obtained replication and metabolic functions, and we are far closer to explaining how the precursors of biomolecules, such as proteins and carbohydrates, were synthesized and assembled from a common starting point.

And while we do not know if extraterrestrial life would rely on the same building blocks that gave rise to life on Earth, we are becoming better equipped to think about the frequency of abiogenesis elsewhere as we approach an understanding of how life arose here.

In the quest to find extraterrestrial life, no question is more important than whether life is broadly deterministic and quite probable or a consequence of random and improbable events. In other words, do the same basic conditions always give rise to life? Or was the emergence of life on Earth a freak occurrence that is very unlikely to happen again?

Numerous fields of study are advancing these questions on all fronts. As they do, one simple observation looms large: the single, substantial source of data we have—namely, Earth—is astoundingly fecund. The factors that allowed life to emerge on Earth, crucial among them our planet's distance from the Sun, didn't result in merely a few microorganisms huddled around hydrothermal vents on ocean floors. They produced a cornucopia of life of such rich complexity that today's flora and fauna rest atop an entire era of reptiles that preceded them. For us to believe that teeming life would be restricted to a single blue marble in the entire expanse of the universe seems the very height of hubris.

Nearly all life on Earth is dependent on the Sun. Not for nothing have humans worshipped it from the dawn of our civilization to the last time you spent an hour lounging on a beach towel. We are literally the stuff of stars; the matter we are made of was produced in the hearts of exploding stars, which then formed planets like the Earth, which then became the material of all terrestrial life, including you and me. And without the warmth and light of the Sun, there would be no plants, no abundant oxygen, no life as we know it.

It is no exaggeration to state the majority of complex multicellular life on Earth is directly or indirectly dependent on the Sun's existence. But what should the search for extraterrestrial life take from this fact? How can we use the known certainty that the Sun supports conscious intellectual life to inform our search for life elsewhere?

Knowing whether or not our Sun is anomalous would tell us a good bit about how anomalous (or not) the life that it supports is. If the Sun is a typical host in all respects, and the presence of sentient life in its vicinity is exceptionally rare, if not unique, then our existence is most likely the result of random chance and unusual indeed. But if the Sun is atypical in certain respects, perhaps those atypical characteristics are required for life, making our existence less random and less unique. That, in turn, would make our search

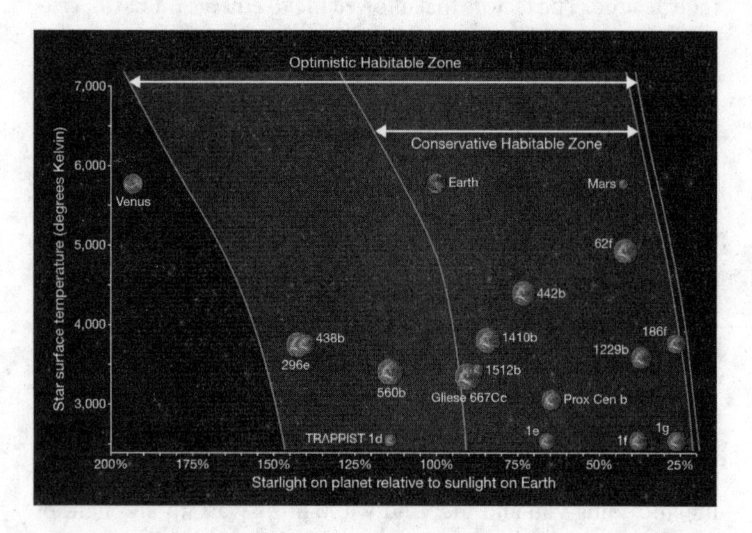

Habitable-zone boundaries around stars with different surface temperatures (vertical axis), ranging from the most abundant dwarf stars, like Proxima Centauri, to rare giant stars, like Eta Carinae. The horizontal axis shows the flux of light shining on the planets' surfaces relative to sunlight on Earth. Various known planets are labeled in the diagram. The nearest habitable planet outside the solar system, Proxima b, appears near the bottom right. Image by Mapping Specialists, Ltd.

for extraterrestrial life less random, for we would have reasons to examine stars like our own.

As it happens, the Sun-Earth system is anomalous in two respects. First, the Sun's mass — 330,000 times that of Earth — makes it more massive than 95 percent of all known stars. Second, the brightness of the Sun allows life to exist on Earth at a safe distance from our star; the habitable region around fainter stars is closer in, where stellar wind can strip the atmosphere of planets and frequent stellar flares can harm life. And while this does not rule out searching for life on planets orbiting more statistically average stars, given that we have limited resources of time and money, it encourages us to look for stars that are especially massive, like the one that sustains us.

The qualities of the Sun encourage us to direct our search for extraterrestrial life — at least initially — to stars that are similar to our own. The qualities of Earth guide our search as well, particularly when it comes to picking which planets to study first.

The observable data from Earth, the one planet that we know supports a dense and complex biosphere, allows us to compose a short list of features we should look for on other planets. But paramount among all the parameters that are essential for Earth's habitability is the existence of liquid water.

Liquid water, often called the universal solvent, is ideally suited to transport energy into and waste out of cells, and no terrestrial life has been discovered that is able to exist without it. It is so important for life, in fact, that astronomers use it to define the habitable zone around each star, measured by a planet's orbital distance from the center of the solar system. Identifying planets at the Goldilocks distance from a star, that zone in which water neither freezes nor evaporates, is the astro-archaeologist's starting point in the hunt for alien civilizations.

The universe, it turns out, provides us with an embarrassment of places in which to look.

Over the past two decades, we have learned that the universe contains numerous exoplanets (the technical term for any planet residing outside the solar system). This spate of discoveries began in 1995, when astronomers Michel Mayor and Didier Queloz became the first to find definitive observational evidence for an exoplanet—a close-in, Jupiter-like planet, 51 Pegasi b, around a sun-like star—based on the line-of-sight motion of this star as the exoplanet orbited it. Their pioneering work ushered in the new era of hunting exoplanets and earned them a Nobel Prize in 2019.

The basic contours of this research were not, in fact, new; they had been put forth four decades earlier by the astronomer Otto Struve, who proposed that the hunt for alien planets might profitably target gas giants zipping around their parent stars in tight orbits, big worlds going around their stars within a few Earth days. The existence of such planets, Struve argued in a 1952 paper, was suggested by the evidence that some binary stars (a pair of stars bound together by gravity) whip around their common center of mass in a similar manner. And these big exoplanets should be relatively easy to detect via their powerful gravitational tugs on their host star or their blocking of light during transits across the star's face.

But Struve's paper was ignored, as was his proposal to search for close-in Jupiters. The scholars that sat on the time-allocation committees for major telescopes asserted that it was commonly understood why Jupiter lies as far from the Sun as it does, and they saw no reason to waste telescope time in searching for exo-Jupiters that were much closer to their host star. Their prejudice slowed down scientific progress by decades.

Once exoplanets were legitimized as part of the mainstream, their discovery accelerated rapidly. Within a decade after the discovery of 51 Pegasi b, hundreds of other exoplanets were identified. And with the 2009 launch of NASA's Kepler Space Telescope, built for the explicit purpose of identifying exoplanets, that number, at the time of this writing, has jumped to 4,284, and thousands of candidates await confirmation. What is more, we now know that about a quarter of all stars are orbited by planets of Earth's size and surface temperature, planets that might have liquid water—and the building blocks of the chemistry of life—on their surfaces.

The abundance of exoplanets upon which we could fix our observational equipment reminds me of a common Jewish tradition during Passover Seder: the hiding of a piece of matzoh, called the afikomen. The task for the children of the household is to find it, and whoever is successful receives a reward.

What I learned as a child—and what I am mindful of now as an adult in the nascent field of astro-archaeology—is that the question "Where to look?" trumps the question "What exactly are we looking for?" And my sisters and I also quickly learned that the best places to start looking were the places where the afikomen had been hidden in the past.

Today, this same strategy is guiding the search for extraterrestrial life. Most of our telescopes and observational instruments are seeking evidence of life on rocky planets with features—most crucially liquid water—that are consistent with the one place we know life exists: Earth.

But is this all we can do? Is there anywhere else we might look, even if we restrict ourselves to the orbits of alien stars?

• • •

Exoplanets that seem similar to Earth aren't the only places where we could seek life. Additional research that I conducted with my postdoc Manasvi Lingam suggests another highly promising place to seek the chemistry of life: in the atmosphere of so-called brown dwarf stars.

Brown dwarfs are small, less than 7 percent of the mass of our Sun. And since they do not have enough mass to sustain the nuclear reactions that cause other stars to burn so bright (and so hot), they can cool to planetary temperatures. This could result in liquid water existing on the surfaces of small, solid particles in the clouds orbiting a brown dwarf.

We needn't stop with brown dwarfs. We should also consider examining green dwarfs, dwarf stars that show the telltale "red edge" in reflected light that is evidence of photosynthesizing plants. By our calculations, green dwarfs orbiting sun-like stars might be our best bet for locating an astrobiological afikomen.

Green dwarfs, brown dwarfs, and exoplanets in the habitable zone of suns—these options in no way exhaust the possibilities for astro-archaeologists, especially when you posit civilizations far more technologically advanced than our own. But at this stage of the search for extraterrestrial life, when our theorizing, observational tools, and efforts at exploration are in their relative infancy, these are the best targets available to us. Outside of our own solar system, that is.

• • •

Even as we contemplate seeking out life in interstellar space, we must admit that we've not exhausted the possibilities within our own solar system; astro-archaeologists should also seek evidence of extraterrestrial life in our planetary backyard.

We could start by looking for technological equipment floating through the solar system. Much as we spotted ʻOumuamua, we might discover—and obtain conclusive evidence about—other artificial objects that originated from other stars. In the first century of our own technological revolution, we sent Voyager 1 and 2 out of the solar system. Who knows how many more such objects an advanced civilization might have launched?

The simplest way to detect passing alien technology is to search under the nearest, largest, brightest lamppost—the Sun. Just as happened with ʻOumuamua, sunlight provides us with valuable information about objects' shape and motion, and it also makes them more visible. We need all the help we can get in this search because, for the moment, our tools for spotting objects like ʻOumuamua are relatively primitive.

As I explained toward the beginning of this book, the telescopes that discovered 'Oumuamua did so accidentally; they were all designed, built, and deployed to accomplish other things. The earliest space archaeologists will likewise have to repurpose existing astronomical tools, at least until such time as the world provides them with instruments explicitly built to their purposes.

In the meantime, perhaps the easiest way for us to look for alien technology in our solar system — and certainly the best opportunity we will have to actually lay hands on it — is to devise a method to detect it as it collides with the Earth. This would require us to find a way to use the Earth's atmosphere to search for artificial meteors. If the object is bigger than a few meters, it could leave behind a remnant meteorite that — if we could detect and track it — might yield the first tangible evidence of extraterrestrial technology.

We can also search the surfaces of the moon and Mars for extraterrestrial technological debris. Whether we liken the moon (which has no atmosphere or geological activity) to a museum, a mailbox, or a dumpster, we can say one thing for sure: it keeps a record of all objects that crashed onto its surface over the past billions of years. Without checking, however, we will never know if it contains the equivalent of a statue, a letter, trash — or nothing.

We needn't restrict ourselves to planet surfaces. Jupiter, for example, could serve as a gravitational fishing net that traps interstellar objects that pass near it. Given scientists' current blinders regarding what's out there, they assume they'll recover only natural rocks or icy bodies like asteroids and comets. And no doubt that is the majority of what we would encounter. But perhaps it is not all we would encounter.

Given the rich reward of such a find, we should make the effort. Yes, it will be vastly more expensive and vastly less certain than low-key surveys, but such an undertaking would be in the same vein as my family walking along a beach examining shells. Perhaps tomorrow's space archaeologists will find the equivalent of an extraterrestrial civilization's plastic bottle.

• • •

The more tools we arm tomorrow's archaeologists with, the farther they can extend their search. As I postulated in collaboration with Ed Turner, from the outskirts of our solar system, one can look for artificial lights that originate from distant cities (or perhaps giant spacecraft). One can distinguish an artificial source of light from an object reflecting sunlight by the way it dims as it recedes from us. A source that produces its own light, like a light bulb, dims inversely with distance squared, whereas a distant object that reflects sunlight dims inversely with distance to the fourth power.

One such tool that space archaeologists could use to great effect is the advanced instrumentation at the Vera C. Rubin Observatory. This wide-field reflecting telescope is expected to commence its survey of the sky in 2022. Along with mapping the Milky Way and measuring weak gravitational lensing to give possible insights into dark energy and dark matter, it is expected to increase humanity's catalog of objects in the solar system by a factor of ten to one hundred. The Vera C. Rubin Observatory is far more sensitive than any other survey telescope, including Pan-STARRS—which, of course, discovered 'Oumuamua.

With our newfound ability to peer farther beyond the solar system than ever before, we could search for artificial light or heat redistribution on the surface of a planet. Breaking out of the constraints of the Drake equation, we could look for technosignatures beyond just communication signals. To see how this might work, consider an exoplanet that is already in our sights.

The tidally-locked planet Proxima b orbits the habitable zone of Proxima Centauri, the star nearest to our Sun. When my colleagues and I worked on the Starshot Initiative, Proxima b was the exoplanet we identified as a possible target for our lightsails. Though Earth-size, the rocky Proxima b always faces its star with the same side. You may recall that my younger daughter pointed out that it would make sense to own two houses on such a planet, one on the permanent-day side where it is always hot and bright and another on the permanent-night side where it is always cold and dark.

But an advanced civilization might find a solution of greater technological sophistication. As I argued in a paper with Manasvi Lingam, the planet's inhabitants could cover the day-side surface with photovoltaic cells that would generate sufficient electricity to illuminate and warm the night side. Were we to focus our instruments on such a planet, the varying level of light from its surface as it moves around its star could tell us if a global engineering project of this type had taken place, and the solar cells on the day side would also produce distinctive reflectance and color. Studies to seek out either phenomenon can be done just by monitoring the planet's light and color as it orbits its host star.

This is just one example of the range of telltale signals that space archaeologists could train their instruments to seek. As our own planet suggests, they could look for evidence of industrial pollution in distant atmospheres. (Indeed, several years before 'Oumuamua appeared in our solar system, I wrote a paper with my undergraduate student Henry Lin and the atmospheric expert Gonzalo Gonzales about searching for industrial pollution in the atmospheres of exoplanets as a signature for advanced civilizations.) And while the atmosphere's contamination by a blanket of pollutants could signal one civilization's failure to outrun the filter, it could signal a different civilization's efforts to intentionally warm up a planet that was otherwise too cold or cool down a planet that was deemed too hot. Astro-archaeological excavations conducted light-years away from their objects of study could also include a search for artificial molecules, such as chlorofluorocarbons (CFCs). Long after a civilization has expired and so ceased to send out deliberate signals, some molecules and surface effects of industrial civilization will still survive.

The sandbox of space archaeology stretches, of course, to the edge of the universe. There is no reason to confine our searches to planets. With that understanding, some scientists could dedicate their efforts to searching for flashes of light from beams sweeping across the sky from great distances. Such beams might indicate a civilization's means of communication or a means of propulsion. When humanity has taken the necessary steps to send lightsail craft out into the universe using the method my team devised for the Starshot Initiative, such technology will result in bright flashes visible to others due to the inevitable leakage of light over the edge of the sails.

In addition, one could search for a swarm of satellites or megastructures that block a significant fraction of the light from distant stars, a hypothetical system known as a Dyson sphere, after Freeman Dyson, the late, great astrophysicist who first envisioned it. Such gigantic megastructures would confront major engineering challenges, and if they exist, they would be rarities. But they also present a possible technological solution to the great filter, and given foresight, means, and opportunity, a civilization facing its own extinction could set out to overcome those challenges. Finding out if such things exist, however, begins with our seeking evidence of them.

The contemplation of such megastructures raises a recurring question that space archaeologists will have to overcome, given that their efforts must presuppose the existence of intelligences greater than their own. That a project like a Dyson sphere strikes humans as overwhelming—indeed, impossible—may simply reflect the fact that we are not yet intelligent enough to undertake it. A civilization far more advanced than our own may well have overcome the hurdles that we, from our more limited understanding, see as insurmountable.

• • •

Two examples of artificial structures from alien intelligence around stars and planets: a Dyson sphere (a hypothetical megastructure constructed around a star to harvest its light) and a swarm of communication satellites around an Earth-like planet.

Pursued to its fullest extent, astro-archaeology would inevitably be humbling—but this is the aspect of it that could yield the richest rewards.

If we can accept that we are very likely less advanced than civilizations that have come before us, this might well lead to our finding ways to speed up our own plodding evolution—a psychological transformation that might allow humanity to leap forward by thousands, millions, or even billions of years.

Evidence is omnipresent for the real probability that humanity has not set the intelligence bar particularly high and that other civilizations have likely cleared it. It is as close as your newspaper, your nearest screen, and your endlessly refreshing newsfeed. The true marker of intelligence is the promotion of one's own well-being, but too often our behavior does the opposite. I have found that paying close attention to the world's most pressing news stories provides ample evidence that we cannot be the smartest species out there.

Humanity has rarely focused on its collective well-being, not over the preceding centuries and not today. Among our current bad habits, we repeatedly opt for short-term benefits over long-term benefits in matters as complex as carbon-neutral energy, as fraught as vaccines, and as obvious as shopping with reusable bags. And for over a century we have been broadcasting our existence to the entire Milky Way in radio waves without pausing to worry if any other civilizations out there might be both smarter than us and more predatory.

Of course, the coordination of effort required to present the universe with a more nuanced, uniform message from human civilization presupposes a civilization capable of unity. Human history gives scant reason to hope for that in the future, at least in the near term.

• • •

For the budding field of space archaeology, another foundational challenge—besides getting the requisite tools and resources—will be adopting approaches that increase our ability to imagine the products of other, more advanced civilizations. To put it another way, we must not let the limitations of our own experience and our resulting presumptions leave us intellectually unprepared to interpret any defunct, discarded, or deliberately sent extraterrestrial technology.

To make the point to my students that we cannot allow the familiar to define what we might discover, I have often used the analogy of cave dwellers discovering a modern cell phone. It aptly applies to the possibility that soon humanity may discover a piece of advanced technological equipment developed by an extraterrestrial intelligence. If we do not prepare ourselves, if we have not allowed for the science of space archaeology, we could act much as these cave dwellers would, imagining the phone to be nothing more than an exotic, shiny rock. And with that shortsightedness, those dwellers would miss the chance to take that million-year leap forward.

One fact is clear. If we assign a zero probability for finding evidence of artificial objects, as some scientists did in the case of 'Oumuamua, if human civilization organizes its efforts and its funding and its scholars by declaring, "It's never aliens," then we guarantee that we will never find evidence of extraterrestrial civilizations. To move forward, we must think outside of the box and avoid prejudice about what we expect to find based on past experience.

As individuals and as a civilization, we also must learn modesty regarding both our potential place in the universe and our potential future in the universe. It is far more statistically likely that we are at the center of the bell curve in universal intelligence than that we define the outer edge of the most intelligent.

Students in my class are often surprised to hear the trivial but sobering remark "Only half of you are better than the median of this class." The same applies to civilizations. That we have discovered many planets similar to Earth and have to date found no conclusive evidence of other civilizations should not lead us to presume our civilization and terrestrial life is uniquely assured of a bright future.

While historians can quibble over whether or not our past suggests a teleological direction toward progress, the universe offers a clear answer: The history of the universe indicates a trend toward extinction—of stars, of planets, of solar systems, and, perhaps, of the universe as we know it. The search for, let alone the discovery of, extraterrestrial technology could jar us from our more limited frame, our habit of looking forward a generation or two and not with the future of our civilization uppermost in mind.

• • •

Let me illustrate the need for a new mindset with a personal anecdote. I made six visits to the same university town in Europe, and my hosts kept putting me in a small hotel room where I would bump my head against the tilted ceiling while taking a shower and had to crawl into my narrow bed without space to stretch my legs. I decided I had had enough. "Next time," I promised myself, "I will reserve a double room." And so I did.

But when I arrived at the hotel on my next trip, the receptionist said: "I see that your wife could not make it . . . I will be glad to downgrade your room reservation to a single room." I said, "No way, please put me in the double room that I reserved." When I mentioned the story to my hosts and asked why space was so limited in this town, they answered, "Because the town has a rule that no building can be taller than the church." This raised my inevitable question: "Why don't you make the church taller?" To which they replied, "Because it has been like that for centuries."

Inertia is a powerful thing. Young people often imagine new worlds, literally and metaphorically, but their revolutionary ideas are frequently met with skepticism and dismissal by the "adults in the room," who lost their enthusiasm for challenging reality in many bruising fights long ago. The "adults" simply got used to things the way they were; they came to accept what was known and to ignore the unknown.

Youth is a matter not of biological age but of attitude. It is what makes one person willing to open up new frontiers of scientific discovery while others try to stay within the traditional borders. Becoming a scientist offers the great privilege of maintaining our childhood curiosity and questioning unjustified notions. But this opportunity does nothing for anyone unless people seize it.

It is commonly believed in the conservative scientific community that intelligent life is probably unique to Earth and that it would be a waste of time and funds to search for artificial signals in the sky or for the debris of dead civilizations in outer space. But this is an ossified way of thinking. Today's new generation of researchers has access to telescopes that could turn this notion on its head. Just as Copernicus revolutionized the prevailing dogma about our place in the universe, our generation can foster a new revolution by "making the church taller" still.

11

'Oumuamua's Wager

IMAGINE LIFE ON OUR PLANET THE DAY AFTER there is irrefutable confirmation of life elsewhere in the universe. Briefly entertain the hypothetical idea that 'Oumuamua had been discovered before October 2017 and that we had the opportunity to launch a spacecraft with a camera that snapped a close-up photo of 'Oumuamua at nearest approach and demonstrated beyond a reasonable doubt that this object was technological debris from an extraterrestrial civilization.

Now ask yourself: What would have followed?

Finding evidence of life on another planet, I believe, would have a profound impact not just on the science of astronomy but on human psychology, philosophy, religion—even education. Whereas at present only a small minority of the scientific community seriously studies the probability of and search for extraterrestrial life, these subjects will become part of our mandatory high-school curriculum the moment we know for a certainty that we are not alone. It is not too much to presume that such a discovery would also affect the way we behave and how we interact with each other, because we might come to feel as though we are a part of a single, unified team, humanity, and stop worrying and warring over mundane issues like geographical borders and separate economies.

Such a discovery would change us in both fundamental and subtle ways—and I have to imagine that most of them would be for the better.

Given the ubiquity of habitable planets, it is the height of arrogance to conclude that we are unique. It is, I believe, the hubris of a young age. When my daughters were toddlers, they believed they were special. But after meeting other children, they developed a new perspective of reality, and they matured.

In order for our civilization to mature, we need to venture into space and seek others. Out there, we might discover that not only are we not the only kids on the block, we are also far from the smartest kids on the block. Just as we once gave up on the belief that the Earth was at the center of the universe, so must we start to act from the clear statistical likelihood that we are not intelligent sentient creatures without peer. Not only are you and I personally going to be intellectually eclipsed by future generations, but humans are the sole creators of a civilization no more and we are very likely a great deal less accomplished than what the universe has already witnessed.

Such a framework of understanding will endow us with a sense of modesty, and modesty will improve our perspective on our place in the universe, which is frailer than we tend to believe. And that could increase the chance we endure. Because with every day that goes by, we are gambling with our civilization's fate—and at the moment, the odds that it will last seem long indeed.

• • •

Think of it as 'Oumuamua's wager—a bet along the lines of the one proposed by the seventeenth-century French mathematician, philosopher, and theologian Blaise Pascal. He framed his famous wager this way: Humans bet with their lives on whether God exists or not. It is better, Pascal argued, to live a life as though God does.

Pascal's reasoning ran as follows: If it turns out that God doesn't exist, you have only given up a few pleasures during your lifetime. If God *does* exist, however, you gain heaven and with it an infinity of rewards. You also avoid the worst of all possible outcomes: an eternity in hell.

In much the same way, I would argue, humanity bets its future on whether ʻOumuamua is extraterrestrial technology or not. And while our wager is thoroughly secular, its implications are no less profound. In a very tangible sense, the promise of betting right, of exploring out among the stars for the life we expect to find there, are the heavens themselves. And, especially when we consider the specter of the great filter and the fact that civilizations with the technological prowess to explore the universe are also highly vulnerable to annihilation by self-inflicted wounds, betting wrong and planning too little and too late could hasten our extinction.

These two wagers are, of course, vastly different in some important ways. For instance, Pascal's wager requires a giant leap of faith. ʻOumuamua's wager requires only a modest leap of hope – specifically, hope for more scientific evidence. That could be something as simple as a single close-up, high-resolution image of an object we have already managed to photograph from afar.

Pascal's eternal cost-benefit analysis required him to posit a divine, omniscient being. To posit that ʻOumuamua is extraterrestrial technology requires only the belief in an intelligence other than our own.

What's more, whereas Pascal had faith and faith alone, we have evidence and reasoning – assets that shorten the odds in favor of ʻOumuamua having been extraterrestrial technology.

There is another reason I find the comparison of these wagers instructive. I have learned that conversations about ʻOumuamua often veer into the religious. I believe this is because we understand that any sufficiently advanced intelligence will appear to us as a good approximation to God.

• • •

"Have your religious beliefs, or beliefs about God, changed in any way in the time you have been studying astronomy?" When a *New Yorker* journalist interviewing me about 'Oumuamua asked me this, I was initially perplexed. Why make the assumption that I am religious? I was, and remain, secular.

But I began to appreciate where this line of questioning had stemmed from during an interview with CNN. Toward the very end of our allotted time, the interviewer asked me, "In our first encounter with extraterrestrial civilization, would we hope they be religious or secular?" Perhaps realizing it allowed for no one-sentence answer, he added that, given the time constraint, I needn't provide my answer.

I think I do need to. And, more important, I believe we need to think more than we have about the wellsprings of this sort of question. 'Oumuamua presents us with an awesome possibility, and we have traditionally struggled with awe.

Over centuries, our civilization has invented means, from myths to the scientific method, to make sense of things that inspire awe in us. And with the passage of time, many such things have moved from the "miraculous" column of human experience over to the mundane. Much of this can be attributed to advances in the sciences. But no discipline of thought is free of the risk of dogmatic blinders, and this is true for scientists as well as theologians.

Consider how a secular person would hear the question posed to me by the CNN interviewer. He might briefly allow that, on the one hand, religious beings are more likely to be ethical – guided by noble values, perhaps, or holding to some injunction that the meek should inherit the universe. After all, most of humanity's religions teach abstract value systems that followers obey, whether out of fear of punishment by a divine entity or out of social interests. A secular person might even grant that a small number of religions, such as Jainism, explicitly espouse nonviolence.

But, this person will point out, even the briefest survey of our religious history should give one pause. To pick an example at random, recall the sixteenth-century Spanish invasion of Central and South America. Fueled by a hatred of idolatry, in 1562 the Roman Catholic priest Diego de Landa Calderón burned in a grand auto-da-fé thousands of Mayan manuscripts, or codices, destroying so many that next to none remain for today's scholars to study. "We found a large number of books in these characters," the priest declared, "and, as they contained nothing in which were not to be seen as superstition and lies of the devil, we burned them all." If we imagine that our first encounter with extraterrestrials or their technology will replicate the Inquisition of the Roman Catholic Church and the practices that followed Hernán Cortés's 1519 arrival in Tenochtitlan, the capital of the Aztec Empire, we would be right to be concerned.

But consider how a religious person is likely to respond when asked, "In our first encounter with extraterrestrial civilization, would we hope they be religious or secular?" Without question, the sciences, including the social sciences such as economics, have steadily increased life expectancy and reduced extreme poverty. But the presumption that we would obviously prefer a civilization that is secular and scientific raises equally well-worn concerns.

Consider a recent century, the twentieth. Both World War I and World War II, among the most lethal conflicts in all of human history, were secular contests over borders, resources, and power. During that same century, eugenics, the supposed science of controlling human propagation for its improvement, lent false credence to racism in the United States and encouraged the Holocaust in Nazi Germany. Then, too, the twentieth century's most boastfully secular experiment, the Soviet Union, frequently demanded that scientific advances conform to Communist ideological tenets. Clearly science, too, is susceptible to orthodoxy, authoritarianism, and even violence.

The difficulty rests, I believe, in the question the interviewer posed. He took the wrong lesson from the evidence supplied by the one civilization available for study, our own. At the scale of an entire civilization, "Religious or secular?" is liable to be a false dichotomy. Judging by human history, both recent and distant, any alien intelligence we encounter is likely to be both religious *and* secular, and that isn't necessarily a cause for concern.

Cast your mind forward once more to the day after we find proof of life elsewhere in the universe. There is one more prediction I feel confident in making: when we learn for certain that we are not alone in the universe, all of humanity's religions—and all of its scientists, even the most conservative—will find ways to accommodate the fact.

My hope is not that the first extraterrestrial intelligence we encounter be either religious or secular but rather that it be animated by humility rather than arrogance. This would qualify the encounter as a mutual learning experience that enriches both parties rather than a zero-sum conflict propelled by self-interests and followed by a power struggle for domination. This hope extends, of course, to the way we should engage in space exploration as we approach distant outposts and as we consider establishing our own settlements—our own little Beit Hanans among the stars. As we proceed farther into the universe, our moral responsibility and humility should follow higher standards than those demonstrated so far on Earth.

In humanity's case, both religion and science have, over the course of our history, bolstered both our humility and our arrogance. It is the height of arrogance to rule out what can be rationally considered, but that is the work of all intellectual blinders, whether they're manufactured by theologians or scientists. Both fields have, on occasion, encouraged their practitioners to don such blinders, restricting their thoughts and forcing them to follow the well-worn traces of existing lines of inquiry.

It must also be admitted, however, that both science and theology have, on occasion, encouraged pockets of their practitioners to do the opposite, to shed their blinders and open themselves up to the new, the controversial, the unexpected. This is where I find cause for hope.

First, it is probable that the members of any extraterrestrial civilization will be as awed to encounter us as we are to encounter them. They, too, will likely have stared out into the abyss of space for countless generations; they, too, will understand that the universe teems with planets capable of supporting life and yet life in the universe seems exceptionally rare.

Second, it is quite possible that they will worry about their reception by our species as much as we may worry over their intentions. Whatever information they have concerning life on Earth, it will be partial and much of it terribly out of date. Just as terrestrial astronomers look back into time when they look out into space, so do extraterrestrial astronomers. The laws of physics, after all, will apply to our alien counterparts' technologies no less than they apply to ours, and based on everything we have learned to date, this suggests that the distances traveled will encourage modesty. Consider that all of humanity's interstellar vehicles are doomed to one-way journeys; the same will likely prove true for the extraterrestrial.

Third, I have to imagine that among the alien intelligence we eventually encounter, there will be a few existentialists. I don't think that is a fantastical concept. Just as the intellectual history of humanity allowed the existential school of thought to flourish on Earth, informing what came after it, I suspect the same will prove true for alien intelligence. I believe that they, no less than us, will have spent a civilization's lifetime confronting life's most stubborn mysteries, the ones that are impossible to move from the miraculous to the mundane.

There is no mystery more fundamental than the meaning of life. Some of us are cast in the role of Hamlet, some of us in the roles of Rosencrantz and Guildenstern, but all of us experience the sensation of striding across a stage without a script. It is the rare human and, I suspect, sentient being who never seeks an answer to the question: What's it all about?

Early in my life I turned to the existential philosophers, especially Albert Camus, for guidance. Among his works, one that resonated with me was *The Myth of Sisyphus.* According to Greek mythology, Sisyphus was punished by the gods and forced for all eternity to roll a heavy boulder up a hill only for it to roll back down once he managed to get it nearly to the top. Camus believed that this was analogous to the absurd condition of man, who was likewise caught in a perpetual cycle as he tried to understand an inexplicable world. Sentient life's common circumstance, to live and die without ever learning why, was, Camus believed, absurd. I believe that other sentient beings—who are bound by intellectual limitations, just as we are—will inevitably arrive at the same conclusion: life is absurd.

It is difficult to remain arrogant in the face of the absurd. Humility is the more apt posture. The more we see evidence of humankind cultivating humility when confronted with the awesome, the more we have reason to anticipate the same from extraterrestrial civilizations.

Over our history, humans have repeatedly sought to fight for causes that were more inspiring than their private lives—usually causes having to do with terrestrial concerns, such as nationality and religion. To take another example chosen at random, during the Second World War, Japanese soldiers were willing to sacrifice their lives for the sake of their emperor, Hirohito. But in view of our recent realization that there are approximately a zetta (or 1,000,000,000,000,000,000,000) of habitable planets in the observable universe, the emperor's status cannot be more significant than that of an ant hugging a single grain of sand on a huge beach.

And what is true of an emperor is no less true for a soldier or anyone else on Earth.

We would do well to look up and look beyond that grain of sand.

Perhaps, rather than behaving like outsize actors in puny roles, we should adopt the perspective of spectators and simply enjoy the dazzling show all around us. And in the broad spirit of stopping to smell the roses (or inspect the seashells), there is much for the spectator to enjoy, both on Earth and off it. If the rich spectacle of events on our planet is insufficiently inspiring, we can use our telescopes to witness an even wider variety of dramas. Within the next decade, the Vera C. Rubin Observatory conducting the Legacy Survey of Space and Time, a ten-year effort to repeatedly photograph more than half of the night sky, will deliver five hundred petabytes of images from our cosmic environment. I dream that one of the survey's results will be a new streaming-media subscription service—this one broadcasting the entire universe.

Not all of us can remain spectators, of course. Some of us will aspire to make a difference. The ways to contribute—and I will always argue that no pursuit holds as much promise of contribution as the sciences—are myriad, but it will help if we give ourselves an objective commensurate with our capacity to wonder, and to hope.

• • •

Pondering the values of other civilizations ultimately allows us to understand and refine our own. The effort also captures the promise of 'Oumuamua's wager.

Bet that humanity has recently brushed with extraterrestrial technology, and we fundamentally shift what we seek and what we expect to find out in the universe. Similarly, we shift those things we could aspire to undertake that would make a difference not just to our world but to the universe at large. Live as if we know there is, or has been, intelligent life in the universe other than our own, and we redefine some of the missions of humanity.

Personally, I have always been driven by the desire to understand something new about the universe that would change our cosmic perspective and stimulate our aspirations for space. I assign a meaning to my life by using the spectator perspective of an astronomer to motivate new challenges for our civilization on the cosmic stage. Given our many engineering accomplishments on Earth, a broader perspective could lead us to develop new technologies, ask new questions, establish new disciplines, and see our role in a larger habitat.

Among all astronomical data, the discovery of alien life could have the greatest impact on our broad outlook. And what if we have already made that discovery? What if, like so many of science's brushes with insights that profoundly change how we see the world around us and our place within it, the only thing preventing us from adopting that broader outlook is our own reluctance to accept optimistically 'Oumuamua's wager?

The main benefit from an encounter with superior beings would be the opportunity to ask them the fundamental question that has been bothering us for ages: What's the meaning of life? I hope to live long enough to be around for their answer, stemming from numerous millennia of acquired scientific knowledge. But at the same time, I worry that the pace of humanity's advance to the answer will be hampered by our arrogance, which has often made it easier for us to cling to our grain of sand rather than look up into the expanse of stars.

12

Seeds

IF, WHEN CONFRONTED WITH 'OUMUAMUA'S WAger, we decide to bet on the object being the product of extraterrestrial intelligence rather than simply a weird rock, that raises another question: How much are we willing to place on this wager?

For starters, consider the least ambitious bet humanity could make: We could simply accept that we missed an opportunity to properly examine the first interstellar traveler that humankind ever spotted and resolve to be better prepared in the future so as not to miss the next one. Preparation could proceed along several tracks, including coming up with a way to capture an image of the next wildly anomalous object that passes through our solar system, perhaps even to capture the object itself. But preparation would also entail increasing every capacity, intellectual and technological, to be able to study and make sense of what we find. The results of such a modest bet are staggering to contemplate—the discovery of another civilization's technology might help us reach goals that we have long aspired to.

Astro-archaeology would be one such initiative—but our efforts shouldn't stop there.

Take seriously the hypothesis of 'Oumuamua's alien technology origin, and we must also take seriously the challenges that we are likely to face in our next encounter with extraterrestrial technology or life. Once we find conclusive evidence of extraterrestrial life in the universe, there will be a predictable international debate about whether and how to react. How do we prepare for this debate? How do we anticipate and plan for the communication SETI has been seeking for decades—or for any other evidence of extraterrestrial intelligence, for that matter?

Bet that 'Oumuamua was an exotic rock and nothing more, and on the day that more such evidence presents itself clearly, we will *scramble* to construct the necessary tools. The first will probably be the academic field of "astro-linguistics" to confront the challenge of an intergalactic means of communication. Other fields would follow, such as "astro-politics," "astro-economics," "astro-sociology," "astro-psychology," and so on.

Bet that 'Oumuamua was of alien technology origin, however, and we could start *tomorrow* to establish such fields.

There are other, relatively unambitious bets we could make on 'Oumuamua's alien technology origins. For instance, upon discovering conclusively that we are not alone in the universe, we will also quickly discover that existing terrestrial international law provides us no thought-out framework to respond to an extraterrestrial encounter. Indeed, among the most modest of all the optimistic bets that humanity could place on 'Oumuamua being of alien technology origin would be the establishment of international protocols and oversight—likely under the umbrella of the United Nations—regarding our efforts to search for and find evidence of extraterrestrial life and communicate with extraterrestrial intelligence. Even a nascent treaty agreed to by all terrestrial signatories would provide a framework for how we, as a species, respond to an encounter with a mature intelligence that is billions of years more advanced than we are.

What would be the most ambitious wager humanity could place on ‘Oumuamua? It would be something sufficient to ensure terrestrial life’s survival.

A more ambitious bet would be to learn from what we imagine a more mature civilization might have attempted. To take the small scientific leap and allow the possibility ‘Oumuamua was extraterrestrial technology is to give humanity the small nudge toward thinking like a civilization that could have left a lightsail buoy for our solar system to run into. It is to nudge us not just to imagine alien spacecraft but to contemplate the construction of our own such craft.

Alien spacecraft might include robots equipped with 3D printers and artificial intelligence, which would allow them to use raw materials they scooped up elsewhere to make artificial objects based on blueprints from their home planet. This serves the purpose of avoiding catastrophes at one location by making copies of the same precious content at other locations. The advantage of 3D printing of life from raw materials on a target planet is that natural biological systems with DNA as we know it live a finite lifetime. Even the most cautiously stored building blocks of life may disintegrate in a few million years. Machinery capable of constructing life once it reaches its destination can last much longer.

Perhaps we should be working along the same lines even before we obtain conclusive evidence that we are not the only or even the most intelligent life in the universe.

As a child, I would search for the spherical seed heads of dandelions, bring them up to my face, and blow as hard as I could. Just as nature intended, the seeds scattered far and wide. Two weeks later, I would see new seedlings pushing up from the soil. Could civilizations preserve themselves from extinction by doing something similar? Might extraterrestrial civilizations have already tried? And could this also be a way to preserve life in the universe?

Recall 'Oumuamua's slight deviation from the path explicable by the gravity of the Sun alone. Something else pushed it, and I have hypothesized that this something else was the force of sunlight on an extraterrestrial lightsail. But even assuming the object to have been optimally designed for this purpose, it deviated only slightly. The reason is that the force of the Sun is barely capable of accelerating a lightsail craft to just a thousandth of the speed of light even if it starts its journey as close as ten times the solar radius (which is the closest we have yet managed to send a craft, the Parker Solar Probe, the robotic spacecraft launched in 2018 to study the corona of the Sun). We would need a far greater force to propel our seeds of terrestrial life in sufficient number and across the universe. Something less like the radiation of our Sun—and more like a star exploding in a supernova.

An exploding star would have a luminosity equivalent to a billion Suns shining for a month. A lightsail weighing less than half a gram per square meter could, propelled by such an explosion, reach almost the speed of light even if it were separated from the exploding star by a hundred times the distance of the Earth from the Sun. This would allow our dandelion craft to reach regions of the universe of which we currently can only dream, thereby dramatically expanding the number of possible planets where the seeds of life might find a home.

To picture how this might work in practice, imagine a civilization that resides near Eta Carinae, a massive star with a luminosity five million times that of the Sun. To ensure the continuation of life, it could park numerous lightsails around the star and cleverly await the explosion that would launch these sails to almost the speed of light at minimal cost.

Such a civilization would have reached a level of either patience or profligacy that humanity hasn't—yet. Massive stars live for millions of years and the exact timing of their explosion is difficult to forecast. Eta Carinae, for example, has a lifetime of a few million years. Predicting its death to the precision of millennia is the equivalent of forecasting anyone's death to within a year after approaching the average life expectancy.

Such a civilization would also have had to plan ahead to a degree humanity has never proven capable. While its lightsails could

The Crab Nebula is the remnant of a supernova explosion observed on Earth in 1054 from a distance of about 6,000 light-years. The remnant contains a neutron star, the Crab Pulsar, near its center that spins 30 times every second and pulsates like a lighthouse. Explosions such as this could be harnessed to propel lightsails to the farthest reaches of the universe. ESO be transported to their destination around the aging star well in advance of the explosion using cheap chemical rockets, the journey could take millions of years using that primitive mode of propulsion.

But it is forethought and patience that present the greatest obstacles. The technology, while formidable, is achievable. We know from our modeling for Starshot that sails must be highly reflective so as not to absorb too much heat and burn up. We can also anticipate how to build such craft to avoid the danger of them being pushed away by the bright starlight prior to the explosion. And to keep these craft from accelerating into paths strewn with stellar debris, they should be designed to fold into needle-like configurations to minimize damage and friction and greatly increase the number of such craft.

It would be a civilization's greatest of hedged bets. Conceivably numbering in the trillions, these small lightsail craft built to preserve the building blocks of life could sit like dormant seeds a tremendous distance from an aging massive star, awaiting the inevitable. Even if the civilization that placed them there failed to outrun its great filter, on going supernova, the star would disperse into the universe the possibility for the continuation of life, just like the seeds of a dandelion do.

Of course, there is no need to be that patient. It is already technologically feasible for humanity to use powerful lasers that will be far more effective than the Sun to push lightsails out into interstellar space. This, of course, is the centerpiece proposal of the Breakthrough Starshot Initiative: a laser beam that produced ten gigawatts of power per square meter would be ten million times brighter than sunlight on Earth and able to send lightsail craft at several tenths the speed of light.

Without question, this would require a major investment. But the moment we know that we are not alone, that we are almost certainly not the most advanced civilization ever to have existed in the cosmos, we will realize that we've spent more funds developing the means to destroy all life on the planet than it would have cost to try to preserve it. Confronted with the 'Oumuamua wager, we might conclude that humanity's continued existence is worth the expense.

• • •

Currently, we keep all our eggs in one basket—the Earth. As a result, humanity and our civilization are extremely vulnerable to catastrophe. By spreading multiple copies of our genetic material through the universe, we could guard against that risk.

This effort would resemble the revolution brought about by the invention of the printing press, which allowed Johannes Gutenberg to mass-produce copies of the Bible and distribute them throughout Europe. As soon as many copies of the book were made, any single copy lost its unique value as a precious entity.

In the same way, as soon as we learn how to produce synthetic life in our laboratories, "Gutenberg DNA printers" could be distributed to make copies of the human genome out of raw materials on the surface of other planets. No single copy would be essential for preserving our species' genetic information; rather, that information would be contained in multiple copies. As I write, colleagues of mine at Harvard and elsewhere are working diligently to move the miracle of creating life into the column of mundane accomplishments. Much as physics benefited greatly from laboratory experiments that unraveled the laws that govern the universe, these scientists are attempting to create synthetic life in the laboratory and unravel the multitude of chemical paths that could give birth to life. For example, the Szostak Laboratory, headed by its namesake, Nobel laureate Jack Szostak, is building a synthetic cellular system that evolves, replicates itself, and preserves genetic information according to the mechanisms outlined by Charles Darwin in 1859. Szostak and his team are focused on creating a protocell capable of replication and variation, which means it should be able to evolve; they hope this will lead to the spontaneous emergence of genomically encoded catalysts and structured molecules.

If it works, this accomplishment would guide us toward the best celestial targets in our astronomical search for life by showing us under which conditions life can emerge. But it might also teach us more about our own selves as life-forms—and give us a much-needed dose of humility in the process.

Consider that cookbooks are full of recipes that have the same ingredients but result in different cakes, depending on the timing and fashion in which these ingredients are mixed and heated. Some cakes taste better than others. There is no reason to expect that terrestrial life, which emerged under random circumstances on Earth, was optimal. There may be other paths leading to better cakes.

The prospect of humanity producing synthetic life in the laboratory also raises interesting questions about our own origins. Are we the result of exclusively terrestrial evolution? Or did we, like the protocells being developed in university laboratories, receive a helping nudge?

• • •

In 1871, in an address before the British Association for the Advancement of Science, the prolific physicist and mathematician Lord Kelvin suggested that life could have come to Earth by way of traveling meteorites.

The idea wasn't original to Lord Kelvin. Ancient Greeks entertained the idea, and decades before Lord Kelvin's address, other European scientists had given the possibility scrutiny. But despite the interest in the nineteenth century, following Kelvin's presentation to the association in 1871 the idea went ignored for another century.

Over the past two decades, however, the theory of panspermia

—the idea that life might reach habitable planets by way of meteorites, comets, or stardust—has been gaining more rigorous attention as scientific research confirmed the hypothesis that certain meteorites discovered on Earth were of Martian origin.

Once we began to seek these Martian meteorites, we found many more of them. We learned that ALH84001, a Martian meteorite that was found in Antarctica in 1984, had never been heated above 40 degrees Celsius (104 degrees Fahrenheit) after being ejected from the surface of Mars. To date, over a hundred such Martian landers have been identified. If there was ever life on the Red Planet, clearly it has had opportunities to reach Earth and survive.

Adding to the intrigue is the fact that scientific consensus has it that Earth was uninhabitable until about four billion years ago, and yet we have found evidence of life dating back 3.8 billion years. How is it possible, scientists asked, for Darwinian evolution to have so quickly produced DNA-based life? We know from terrestrial biology that life is self-serving. Selective and spontaneous adaptations that increase life's ability to persevere is the bedrock of Darwinian biology. Life's aim is survival, which means propagation. How plausible is it that life would make use of panspermia to spread and secure its survival?

In 2018, my postdocs Idan Ginsburg and Manasvi Lingam and I published a paper entitled "Galactic Panspermia" in which we presented an analytical model to estimate the total number of rocky or icy objects that could be captured by planetary systems within the Milky Way galaxy and result in panspermia should they harbor life.

We first considered whether we might be Martians. For life on Earth to be descended from life on Mars, the Red Planet would have to have been hit by an asteroid or comet with sufficient force that material was ejected into interplanetary space, and that material would have had to find its way to Earth. And, crucially, any life aboard would have had to survive the interplanetary voyage as well as the ejection and landing.

Mars has indeed been struck trillions of times by space debris larger than a person over the billions of years of its existence. Many impacts have resulted in temperatures and shock pressures guaranteed to kill any building blocks of life still clinging to the ejected rocks. But, as with the Martian meteorite ALH84001, some ejected materials did not exceed the boiling temperature of water, allowing some microorganisms to survive. This means that were there Martian life, it would still be living on rocks tossed into space by these gentler impacts. Scientists estimate that Mars has ejected billions of such fragments—objects with low enough temperatures to allow life to survive.

But even if the microorganisms survive their ejection from Mars, how plausible is it that they could survive the trip? There has been lively debate on this point, especially over how lethal ultraviolet radiation is to bacteria. But radioresistant bacteria with extreme tolerance to ultraviolet radiation and ionization have been discovered, and these strains could survive such a trip. (In fact, some terrestrial bacteria exhibit such extreme tolerances to UV and radiation that it is likely they originated on Mars.) What's more, the hypothetical pool of surviving bacteria increases if we posit their traveling within a meteorite or comet in a way that shields them from UV radiation; such a rocky shielding could be as thin as just a few centimeters. And other studies have proven that spores of the bacteria *Bacillus subtilis* can survive in space for up to six years; other bacteria could live for vastly longer stretches of time, measurable in millions of years. Further, scientists have hypothesized a colony of bacteria that's able to surround itself with a biofilm that greatly increases the organisms' protection from harmful radiation.

In another paper, my undergraduate student Amir Siraj and I calculated that bacteria floating in the Earth's atmosphere could have been scooped up by grazing objects that passed just fifty kilometers above sea level and then escaped the solar system. Such an interstellar-space-bound object would resemble a spoon passing through the foam on top of a cappuccino, only in this instance it would continue on with a residue of terrestrial life. We found that billions of such "spoons" have stirred the Earth's atmosphere over the planet's lifetime.

Would the bacteria have survived the trip? It is well known that fighter pilots can barely survive maneuvers with accelerations exceeding ten gs, where *g* is the gravitational acceleration that binds us to Earth. But Earth-grazing objects would scoop up microbes at accelerations of *millions* of gs. Could they live through the jolt? Possibly. Microbes such as *Bacillus subtilis, Caenorhabditis elegans, Deinococcus radiodurans, Escherichia coli,* and *Paracoccus denitrificans* have been shown to live through accelerations just one order of magnitude smaller. As it turns out, these mini-astronauts are far better suited for taking a space ride than our very best human pilots. They could survive the impact on Earth's surface as long as their deep interiors were not overheated, similar to the Martian rock ALH84001.

This data tells us we cannot dismiss the possibility that we are of Martian origin. But might we be even more exotic? Might the truly original source of life on Earth, whether or not by way of a Martian layover, be interstellar or intergalactic? Yes. After conducting a rigorous analysis of the viability of panspermia, my colleagues and I determined that there is a parameter space that allows the galaxy to be saturated with life-bearing objects. While objects with lower velocities are more likely to be captured by the gravitational pull of a planet, and given the established fact that some bacteria can survive millions of years, we estimated that the probability of a life-bearing object striking a planet was significant. Indeed, by positing gravitational scattering events at the galactic center of the Milky Way, we projected that rocky material could be ejected at such extreme velocities that the center could have seeded the entire galaxy.

Those seeds need not even be restricted to bacteria. Certain viruses, which are also capable of Darwinian evolution, have proven themselves sufficiently durable. Even more complicated life might make the trip. Indeed, two roundworms discovered in the Arctic permafrost were revived after being in cryobiosis—when metabolic processes stop—for an estimated thirty to forty thousand years. If organisms such as these could survive the kinds of conditions and timescales that they might encounter on an interplanetary voyage, who's to say they themselves aren't descended from Martians?

Here is where the right bet on 'Oumuamua's wager could pay immediate dividends. Gamble that we have already seen evidence of extraterrestrial intelligence, and both the questions we ask and the projects we undertake shift. Consider that every scientific contortion we've just taken to arrive at the higher probability of naturally occurring panspermia is simplified if we entertain *directed* panspermia. How to ensure life is safely ejected from a planet? Eject it yourself. How to ensure life is sufficiently protected from the harms of space while traveling between planets or galaxies? Build a rocket to the purpose. How to make sure life is nurtured and preserved to survive the extremely long journeys among galaxies? Build your rocket to that purpose too.

• • •

Much depends on how we respond to 'Oumuamua's wager. The safest bet is to deem the object a peculiar rock, nothing more, and stick to our familiar habits of thought. But when so much is at stake, safe bets can only get us so far.

If we dare to wager that 'Oumuamua was a piece of advanced extraterrestrial technology, we stand only to gain. Whether it prompts us to methodically search the universe for signs of life or to undertake more ambitious technological projects, placing an optimistic bet could have a transformative effect on our civilization. If humanity is able to think, plan, and build in pursuit of a vision measurable in millions of years, we just might manage to ensure that life in the universe is able to ride out the vast challenges of time and space by riding the flash of light from an exploding star. When I think of this familiar technology in that way, a lightsail tumbling in sunlight resembles nothing so much as the wings of a dandelion seed sent off by the wind to fertilize virgin soil.

Which brings us back to life originating in laboratories. Take the more cautious approach to 'Oumuamua's wager and we celebrate this extraordinary accomplishment solely for its implications for biomedical research. Take the more ambitious approach to 'Oumuamua's wager, and creating synthetic life in the laboratory becomes, potentially, a means for terrestrial life to outrun the great filter, even after the inevitable death of the Sun.

There is no doubt that if our civilization is bold enough and lasts long enough, we will eventually migrate into space—and into new regions of the universe that, in essential ways, resemble our current one. In doing so, we will surely be following in the footsteps of those who came before us; just as ancient civilizations migrated toward banks of rivers on Earth, advanced technological civilizations likely migrate throughout the universe toward environments that are rich in resources, from habitable planets to clusters of galaxies.

But no civilizations, very much including our own, will make the leap to migrating out among the stars if they are not smart enough to preserve their home planets while they plan and prepare. And it is an achievement humanity is less likely to attain when so many of us cling to the uniqueness of terrestrial life like that ant clung to that grain of sand.

13

Singularities

'OUMUAMUA IS EXTRATERRESTRIAL TECHNOLOGI-cal equipment.

That is a hypothesis, not a statement of fact. Like all scientific hypotheses, it awaits its confrontation with data. And as often happens in science, the data we have is not conclusive, but it is substantial.

Is there any chance that we could obtain additional data about 'Oumuamua or similar objects beyond what we have already gleaned?

The last time we saw 'Oumuamua, it was moving away from us incredibly quickly—many times faster than our fastest rocket. But of course, we could develop spacefaring technologies that are faster than rockets, like lightsails. Or we could approach the next 'Oumuamua-like object with conventional rockets as it's on its way toward us.

If we were to launch a spacecraft close to such an object, we might be able to photograph its surface. What evidence might we find? Almost all of it would be refinements on what we currently know. The right sorts of imaging would yield more data as to its size, its shape, its composition, its luminosity, perhaps even tell us if it bears obvious markers of its manufacturers, just as NASA always stamps its rockets with the American flag. Whatever the evidence is, I would welcome it.

• • •

Unless and until we obtain additional evidence about 'Oumuamua-like objects, we need to work with what we have. And what we have can be summed up with one repeating theme:

And yet it deviated.

'Oumuamua, a small interstellar object first discovered by humans on October 19, 2017, that was highly luminous, oddly tumbling, and most likely disk-shaped, deviated from a path explicable by the Sun's gravity alone without any visible outgassing. All of its properties, very much including its origin in space-time being local standard of rest, rendered it a statistical outlier to a highly significant degree. As a member of a population of objects on random orbits, it required much more solid material to be expelled than available in planetary systems around other stars. But if 'Oumuamua was extremely thin or its orbit was not random, the problem could be alleviated.

Overwhelmingly, the scientific community has coalesced around the conclusion that 'Oumuamua was a naturally occurring object, a peculiar, even exotic comet, but still, for all its peculiarities, just an interstellar rock. And yet it deviated.

It is true that we can hypothesize natural phenomena that could explain each of 'Oumuamua's observed exotic features. There is a statistical possibility, roughly one in one trillion, give or take, that 'Oumuamua was a unique rock. But then ejecting enough material from planetary systems around nearby stars to supply a random population of 'Oumuamua-like objects becomes even more challenging, because now one needs much more material in the form of normal interstellar objects, like 2I/Borisov.

Alternatively, the data allows another hypothesis: that 'Oumuamua was extraterrestrial technology, perhaps defunct or discarded. In that data is something underappreciated by nearly everyone who has written on the subject. It is the fact that humanity could build within a mere few years a spacecraft that would demonstrate every single one of 'Oumuamua's features. In other words, the simplest, most direct line from an object with all of 'Oumuamua's observed qualities to an explanation for them is that it was manufactured.

The reason most of the scientific community cannot keep comfortable company with this hypothesis is that we didn't manufacture it. To allow the possibility that another civilization did is to allow the possibility that one of the most profound discoveries—that we are not the only intelligence in the universe—has just passed through our solar system. It forces us into a new way of thinking.

• • •

Accepting my hypothesis about 'Oumuamua requires, above all else, humility, because it requires us to accept that while we may be extraordinary, in all likelihood we are not unique.

When I say we are extraordinary, I don't mean it literally. That we are the stuff of stars is a poetic truism, but less poetically, it can also be said that the stars are made of the same stuff as us. That goes for the universe as well, for all that is within it started in the same dense soup of matter and radiation emanating from the Big Bang. Still, as I tell the students in my freshman seminar, although all of us are composed of the same ordinary matter, that does not preclude us from becoming extraordinary *people.* Far more significant is that the organization of the stuff we are made of has, over the course of millennia, become the stuff of life. And unlike everything we have discovered in the universe to date, we and we alone are so organized.

Extraordinary and *unique* are importantly different things. Consider Nicolaus Copernicus, the sixteenth-century astronomer who first proposed that the planets orbit the Sun and, in so doing, made a supposedly unique contribution to our conception of the cosmos. His book asserting this thesis was published shortly before his death in 1543, and it went largely ignored by all but a small number of astronomers, most of whom were Nicolaus's friends. But today we trace the origins of a heliocentric solar system to Copernicus, and we use his name to describe the principle that neither Earth nor humanity occupies a special part of the universe, and, indeed, that the universe has no unique or special places. It is the same here, where humanity exists, as it is everywhere else. Today, we can add an ironic codicil to the Copernican principle: there is nothing special about a species and a civilization that has figured out this fundamental fact about the cosmos, for the same thing has likely been figured out by all civilizations everywhere else in the universe.

If we don't merely entertain this thought but embrace it, we find ourselves confronting amazing possibilities.

When Matias Zaldarriaga and I realized that human civilization produced a great deal of noise at the meter-wave radio spectrum, we thought it reasonable that another civilization might produce noise in the same radio band, and so we proposed seeking evidence of that. When Ed Turner and I realized that Tokyo would be visible through a Hubble Space Telescope placed out at the edge of our solar system, we thought it reasonable to seek a similar glimmer from another civilization's city or spacecraft. Similarly, when my postdoc James Guillochon and I realized that humankind could feasibly send out lightsail-propelled spacecraft, we knew that it also stood to reason that another civilization could come to the same realization—and so we recommended a search for the telltale beams of radiation from such launches.

In that same spirit, it is reasonable to imagine that any such effort by another civilization to send a lightsail craft would have been preceded by something roughly equivalent to the launch of the Starshot Initiative—the project that we undertook to engineer (if not yet actually build) lightsails of our own.

I like to imagine I now know what they went through to get there.

I imagine the pacifists among them would have worried that a spaceship powered by a 100-gigawatt laser hurtling toward an alien civilization at a fraction of the speed of light could well be interpreted by them as threatening or even as a declaration of war. To which the chair of the advisory board of their version of the Starshot Initiative would likely have answered, as I did, that such a risk was infinitesimal. For starters, I said, we have no knowledge of the existence, let alone the nature, of extraterrestrial life, intelligent or not. If other beings do exist, our craft of only a few grams is unlikely to be noticed, and carrying the energy of a common asteroid, it would easily be classified as such. And it is an utter impracticality to try to aim our small craft to hit a planet light-years away. This would require an angular precision of a billionth of a radian and there is no way for us to know the relative positions of the planet and the spacecraft to that precision over the decades-long journey. No, rather than target a planet, the craft would aspire to approach an orbital region thousands of times larger than the size of a planet, implying a chance impact probability of less than one part in a million.

I can imagine their engineers doubting the project's feasibility. What of damage to the craft due to the impact of interstellar dust grains or atoms? Their board members likely would have nodded, as I did, and noted that a coating of just a few millimeters would prove sufficient to protect the craft and its cameras. The more optimistic of their engineers might well have bemoaned the absence of any deceleration mechanism, and it likely would have been politely pointed out that this was a built-in constraint. Given the distances, the necessary minimal weight of the craft, the speeds at which it had to travel, taking photographs during fly-bys was a sufficiently grand ambition. And *grand ambition* sums it up well. Perhaps these photographs would let us know if there was vegetation there, or an ocean, or even some signature of civilization, all things we would want to see from up close rather than at the distance of our most powerful telescopes.

I would be willing to bet that as these scientists made the case for the project, they would have confronted fiscal conservatives who doubted if such an undertaking was worth the price tag. And I imagine in turn the board behind the effort would have pointed out, just as the Starshot board did, that the economies of scale were actually stunning. In our case, I said that, yes, the construction of the laser would be expensive. And, yes, getting the lightsail craft above the planet's atmosphere would be costly, but the craft themselves would be cheap; the StarChips would cost only in the range of hundreds of dollars each. Which meant that once the costly investments had been made, it would be perfectly reasonable to launch one every several days, aiming them at many hundreds, if not thousands, of targets.

And then, I hope, the optimists among my distant counterparts, armed with scientific knowledge and accompanying humility, would have pointed out that for all the limitations and risks, launching these lightsail craft represented the next giant leap forward. And enough of these alien scientists would have stared out at their stars, just as we stare out at ours, that—awed by the scale of the universe when placed against the scale of their planet, even their solar system—they would have blessed the effort. They would have concluded that these lightsails were the next, best feasible step toward reaching the stars. And, perhaps, they would have allowed themselves to imagine, just as we are doing, that their fast-moving, peculiar-shaped lightsail craft would someday be seen and understood as an announcement and an invitation: "Welcome to the interstellar club."

• • •

It requires imagination as well as humility to acknowledge the utter ordinariness of humanity. Both qualities, I believe, are integral to our ability to outrun the great filter. But so is another: our willingness to entertain the simplest explanation for 'Oumuamua's properties—that they reflect designed intent, not complex accident.

Earlier in this book I invoked William of Occam and his famed razor—that is, the injunction that the simplest solution is likely the right one. Whether confronting 'Oumuamua or any phenomenon, we would be well advised to pick it up. It is a razor that, I have found, struggles to shave an arrogant chin.

Alas, simplicity is not always in vogue.

"Should we make our theoretical model more complex so that our explanation of the data will not appear too trivial?" The question came up in a meeting with my postdocs as they described their projects, a few of which were nearing completion. I was initially surprised and then, as they explained their reasoning, sobered.

The virtue of simplicity should be obvious, especially to astronomers. After all, the power of Copernicus's heliocentric explanation of the solar system was its simplicity; the prevailing theory he helped overturn, Greek astronomer Ptolemy's Earth-centered planetary system, demanded ever more torturous contortions the more evidence accumulated. The failure of Ptolemy and the success of Copernicus remains among the most cited to budding astronomers. Their task, centuries of instructors have explained to students, is to seek the simplest explanation to the data and avoid the hubris of the Greek polymath Aristotle, who, for all his genius, was driven by his need for perfection in the universe to declare, despite the evidence, that planets and stars could move only in perfect circles. His error became unquestioned fact for centuries.

Similarly, for the last decades of the twentieth century, astrophysicists were skeptical of a model of the early universe that was characterized by a small number of parameters—in a word, *simplicity*. The data was scarce, and the majority of astrophysicists concluded the model was surely naive. But by the beginning of the twenty-first century, enough data had been collected to prove that the universe did indeed start from the simplest possible initial state. The early universe, the data showed, was nearly homogeneous (the same everywhere) and isotropic (the same in all directions), and the complex structures we find in it today can be explained by an unstable gravitational growth of small, primordial deviations from these ideal conditions. This simple model is now the foundation of modern cosmology.

Given all these cautionary tales, it might seem incomprehensible that a group of Harvard postdocs in the early twenty-first century were wondering aloud if they ought to *add* complexity to their work. But in fairness, they had good reasons.

In today's fierce job market, the single greatest imperative appears to be impressing one's senior colleagues. The junior scholar can feel it necessary to produce lengthy derivations marked by challenging mathematical complexity. As one postdoc said to me, "I am facing the strategic dilemma of choosing between two options for my future career: long complicated projects or short insightful papers."

In many cases, senior scholars wish to make their work nuanced and less accessible to scrutiny. They have learned that sophistication is valued as a trademark of the elite, and many are rewarded accordingly.

In my research and my mentorship, I try to offer my junior colleagues a counterexample. I tell my own postdocs that accessible short insights tend to stimulate the field, encouraging follow-up work by the scientific community; I urge them to believe, as I do, that brief, intellectually rich work will improve their job prospects; and I tell them that the ability to explain research clearly depends on their describing only those things that they understand and admitting those things they do not. But they inevitably respond: That's easy for you, chair of the Harvard astronomy department, to say.

This is a dilemma indeed, and I fear the effect it will have on science in the twenty-first century—and not just within the scientific community. In academia, rewarding complexity for complexity's sake directs talent and resources in some directions and not others. It can also encourage the isolation of scholarship among a self-identified elite, leading them to disregard the interests of the public that substantially fund their efforts.

This is a serious problem with consequences that reach far beyond the academy. To understand why this is the case, consider one of the greatest mysteries confronting astrophysicists today: the science of black holes.

• • •

Within weeks of our announcing the Starshot Initiative, in April of 2016, I inaugurated Harvard's Black Hole Initiative, or BHI—the world's first center for the interdisciplinary study of black holes. The timing of the two events were close enough that after Stephen Hawking appeared alongside me, Yuri Milner, and Freeman Dyson in New York City, he was able to join me and my colleagues in Cambridge, Massachusetts, to announce the goals of BHI.

It was fortuitous that Stephen could be involved, and BHI's launch was auspicious for another reason: a hundred years earlier, Karl Schwarzschild, the German astronomer and physicist, solved

Albert Einstein's equations for general relativity, a solution that described black holes decades before there was any astronomical evidence that they existed. And a hundred years on, astronomers still hadn't managed to photograph one.

BHI's inaugural event was memorable for many reasons. For one thing, the launch of this historic project was a sought-after professional objective for me—one more matchbox in which to collect promising matches. For another, BHI represented an interdisciplinary approach to science that I have long advocated, bringing together under one roof astronomers, mathematicians, physicists, and philosophers.

But there were simpler satisfactions as well. At the launch event, a photographer was on hand, and in one picture my younger daughter, Lotem, joined Stephen Hawking and my colleagues on stage. It wasn't planned, but in hindsight I think her presence was essential. Scientific advances are cross-generational efforts, and the benefits of human progress accumulate over centuries. Think of the many thousands of telescopes that now dot the planet and the few that orbit it, all of which stand in lineal descent from the one Galileo used on the same sky.

Later, my wife, daughters, and I hosted Stephen and a small number of colleagues at our house for Passover dinner. Of all the speeches that were given over the course of days BHI was announced to the world, the most meaningful to me was the short, several-minute one Stephen gave at my home. Speaking to a small group assembled in our living room, he drew our attention back to the Starshot Initiative and out into the cosmos. "It's been a busy trip," he said.

> Last week in New York, Avi and I announced a new initiative that is about our future in interstellar space. Breakthrough Starshot will attempt to build a spacecraft that can reach twenty percent of the speed of light. At that speed, my trip from London would have taken less than a quarter of a second

(though longer, if you count customs at JFK). The technology that Breakthrough Starshot will develop—light beams, lightsails, and the lightest spacecraft ever built—could get to Alpha Centauri just twenty years after launch. Up to now, we have only been able to observe the stars from a distance. Now, for the first time, we can reach them.

Stephen's words stayed with me, especially because this would prove to be his final visit to the United States. He had told our little gathering, "I hope to return soon in support of the new Black Hole Institute," but he passed away less than two years later, never to witness the project's success or the interstellar exploration of which he had dreamed.

Another set of remarks from around this time has also stayed with me—but for less happy reasons. At that first BHI conference, a philosopher concluded his talk by stating that "conversations with some prominent theoretical physicists led me to conclude that if the physics community agrees on a research program for over a decade, then it must be correct." My prompt skepticism brought to mind a single word—or, actually, a name: Galileo.

Galileo is supposed to have declared after looking through his telescope, "In the sciences, the authority of a thousand is not worth as much as the humble reasoning of a single individual." Einstein, centuries later, got at the same idea when twenty-eight scholars contributed essays to a 1931 book titled *A Hundred Authors Against Einstein* that declared his theory of general relativity wrong. He is supposed to have replied that if he were wrong, then one author with conclusive evidence to disprove the theory would have been sufficient.

A guiding premise of the Black Hole Initiative is valuing the conflicting insights arising from the reasoning of many individuals approaching problems from differing vantage points. That participants were all interested in slightly different things was a strength. The astronomers were hoping to get the first image of a black hole;

the physicists were focused on solving an apparent paradox about how the laws of physics are affected by black holes; and the mathematicians and philosophers were working to figure out the nature and stability of the singularity at the center of a black hole. (The philosophers, in particular, were an essential part of this team, for an honest philosopher constitutes the canary in the coal mine, warning the gathering if intellectual honesty is violated.)

If there was a common denominator at BHI, it was our shared excitement over seeking data to better explore the unexplained anomalies and questions about black holes. And what challenges they are. Here's a short list.

A major anomaly about black holes is what scientists call "the information paradox": Quantum mechanics states that information is always preserved, and yet black holes can absorb information and then evaporate into purely thermal blackbody (information-free) radiation, a phenomenon demonstrated by Stephen Hawking. Do the laws of physics break down at the edge of black holes or is something else going on?

Another major anomaly about black holes is the fact that they seem to "disappear" matter. Where does the matter pulled into a black hole go? Does it collect into a dense object at the center of the black hole, or does it exit our universe and emerge in another one, like water draining into a distant reservoir?

But more generally, might black holes provide insights that would direct us toward unifying general relativity and quantum mechanics? On his deathbed, Einstein sketched his last thoughts on this theory but did not resolve the tremendous challenge. Stephen Hawking similarly spent his final years considering whether properties of black holes would resolve the challenge. While neither man's considerable intellect was sufficient to crack the problem, many astrophysicists and cosmologists are working in their wake.

Finally, one question bothering astronomers at the time of the BHI's founding was less an anomaly than a glaring gap in our evidence. While we had decades of data confirming the existence and properties of black holes, we had never managed to photograph one.

That changed in 2019. The story of how it happened—how the very first photo of a black hole was taken and how humanity was able to obtain such a crucial piece of evidence in our ongoing investigation of this cosmic mystery—is a wonderful illustrative example of how the deliberate, collaborative pursuit of evidence can accomplish the previously unaccomplished. For those of us who don't consider 'Oumuamua to be a closed case, who hope it proves sufficient provocation for humanity to bet on the more ambitious projects, the story of this incredible accomplishment is also a reminder that when humanity works together, we can achieve the unimaginable—feats of research, discovery, and technological innovation that under other circumstances would have been impossible. For instance, constructing a telescope the size of Earth.

• • •

In a 2009 article for *Scientific American* that I coauthored with my former postdoc Avery Broderick, we called the challenge "shooting the beast." For starters, there is the distance. Sagittarius A* is the closest supermassive black hole to Earth, twenty-six thousand light-years away. Another target we were first to recommend, in a dedicated paper published in the *Astrophysical Journal* that year, was the black hole that was eventually photographed, M87, which is fifty-three million light-years away but is substantially bigger. Still, at that distance, photographing it was likened to trying to capture an image of an orange on the surface of the moon.

Hence the need for a really big telescope. More accurately, what was needed was an Earth-size interferometer formed by linking radio dishes across the face of the planet. Accomplishing this required the collaboration of many sites around the world, an observational effort led by my BHI collaborator Shep Doeleman. It resulted in what was dubbed the Event Horizon Telescope (EHT).

Astrophysical black holes, by definition, do not emit light on their own. In fact, they do the opposite—they consume light, along with everything else. But the matter, typically gas, that swirls around them does emit light as it heats up under the stress of the black hole's gravity. Some of that light escapes the pull of gravity, and some is absorbed by the black hole, and the result is a silhouette surrounded by a ring of light that delineates the region around the black hole from which light cannot escape. This is the defining feature of a black hole, its event horizon, or the spherical boundary at which material flows only one way. It is the ultimate prison—you can get in but you can never get out. Astrophysical black holes are hidden behind event horizons, so, just as in Las Vegas, what happens inside the horizon, stays inside the horizon. No information leaks out.

And that is what the EHT was seeking to do: directly observe a black hole and photograph its silhouette. The mission was years in the making. The Black Hole Initiative helped process the data that produced the images that, for weeks in April of 2019, were ubiquitous, and not just within the halls of academia. This globe-spanning effort, requiring a globe-spanning telescope, produced a photograph that galvanized the imagination of humanity. A decade earlier, Broderick and I had sketched what we thought might be the result for the black hole in the giant galaxy M87, and it was particularly rewarding to see a real image of a black hole resembling those sketches appear on the front pages of major newspapers and magazines.

There is a clear link between this success and my work on SETI. An explicit goal of the Black Hole Initiative is to spark interest not only across academic disciplines but also in the general public. We want—indeed, need—to capture the public's imagination. We need our detective stories read, need our efforts to match theory with the rigorous contact with data sufficiently understood so that all of humanity can celebrate scientific successes. Only in this way will we cultivate as many bright, aspiring minds as we need to to meet the current challenges and those of the future.

Then, too, scientists *owe* the public—literally. We are funded by the public. In large measure, most scientific advances can be traced back to government grants paid for by public taxes. Thus, every scientist who has directly or indirectly benefited (which means pretty much all of us) bears a burden to explain not only the work but also the methods used in that work. We have an obligation to report back about our most exciting discoveries and conjectures on topics that resonate with the public, such as humanity's cosmic origins, black holes, and the search for extraterrestrial life.

Science is not an occupation of the elite in isolated ivory towers but an endeavor that benefits and excites all humans, irrespective of their academic backgrounds. I believe this is especially true when viewed from the vantage point of astrophysicists. The questions the universe presents us with are awesome and galvanizing. They are also humbling. Our job is to stare out at events that occurred long before we arrived and objects that will exist long after we're gone. Compared to the subjects of our study, we have the briefest of windows available to us, precious little time to study the universe and attempt to tease out the answers to its mysteries and paradoxes.

• • •

I put my faith and hope in science. Throughout my life, my optimism has provided immediate rewards. Indeed, this experience of getting something for nothing, rich rewards in return for the simple, humble practice of the detective work of science, brings me to a closing thought.

With Paul Chesler, a postdoctoral fellow at Harvard's Black Hole Initiative, I theorized the fate of matter as it approaches the singularity of a black hole. We decided to address the question by way of a simple theoretical model that combined quantum mechanics and gravity. And as we examined the mathematical implications of the model, we realized that it also applied to the time-reversed problem, in which matter expands rather than contracts. This suggested we needn't run the risks of a trip into a black hole, which was likely to rip us apart by gravitational tide and certain to preclude Facebook posts, but could, instead, observe at no risk the *expanding* universe. That is, we could look up and around us at all the matter that started from an initial singularity in time, the Big Bang. The same equations that described a black hole singularity, we realized, could be used to figure out how the universe had obtained its accelerated expansion.

Just like the biblical story of Saul finding his kingdom by chance while searching for his father's lost donkeys, Paul and I stumbled on an unexpected insight while pursuing a completely different goal. By aiming to better understand black holes, we uncovered a mechanism for explaining our accelerating universe.

Our theoretical model is incomplete. It requires much fine-tuning. Even if the model stands up to theoretical scrutiny, it will need to make new predictions that survive the guillotine of future data. Some or all of that work may prove useful for other theories and in other corners of science. And in the aftermath of 'Oumuamua's visit, it leaves me with a thought that never ceases to haunt me. It, too, is a lesson I have taken from our interstellar visitor.

An encounter with another civilization, as I have said, may be humbling. And given all that we might learn from an advanced civilization, in particular, we should even *hope* to be humbled. Such a civilization will no doubt know the answers to a great many questions we haven't figured out and perhaps haven't even asked. But in order for us to gain some intellectual credibility, it would be nice to start the conversation by offering our own scientific wisdom about how the universe was born.

Conclusion

MANY SCIENTISTS ARGUE THAT WE SHOULD COMmunicate information to the public only once our collective detective work has produced a nearly unanimous conclusion. These colleagues of mine believe that discretion is necessary to preserve our good image. Otherwise, they reason, the public could come to doubt scientists and the scientific process. Indeed, this occurs even when there *is* near unanimous conclusion among scientists. Consider, they often point out, the minority of the public who still question climate change. Stepping into controversy that could erode the stature of science, they worry, is too great a risk.

But my view is different. I think that in order to be credible, we need to show that scientific inquiry is a process that is more common and familiar than much of the public presumes. Too often, the approach taken by my colleagues contributes to the populist view of science as an occupation of the elite and fosters a sense of alienation between scientists and the public. But science is not some ivory-tower affair that, through inaccessible means, yields ironclad truths that can be dispensed only by sages. The scientific method is, in fact, closer to the commonsense approach to problem solving that a plumber adopts when trying to fix a leaking pipe.

Indeed, I think researchers and the public would benefit from viewing the practice of science as not so very different from the practices of a wide swath of other professionals. We confront confounding data just like a plumber does a blocked-up pipe, and we draw on our knowledge, experience, and colleagues' wisdom to proffer hypotheses. And these we test out against the evidence.

The outcome of the scientific process is not up to the practitioners, since reality is determined by nature. Scientists are just trying to figure out what reality is by collecting as much evidence as possible and arguing about various interpretations when the evidence is limited. This reminds me of what Michelangelo said when he was asked how he produced such beautiful sculptures from a block of marble: "The sculpture is already complete within the marble block, before I start my work. It is already there, I just have to chisel away the superfluous material." Similarly, scientific progress is about collecting evidence that allows us to remove the large number of possible hypotheses that are superfluous.

The experience of having to reject some of our false ideas is humbling. We should not take our mistakes as insults but rather as opportunities to learn something new. After all, our modest island of knowledge is surrounded by a vast ocean of ignorance, and only evidence—not unwarranted convictions—can increase the landmass of this island. Astronomers especially should be inspired to modesty. We are forced to confront our insignificance in the cosmic scheme of the universe, and against the vast expanse of all physical phenomena, how limited is our understanding. We should be humble in our approach, allow ourselves to make public mistakes and take transparent risks as we attempt to learn about the universe. Just like children.

As I watched colleagues close ranks against the serious consideration of the hypothesis that 'Oumuamua could be extraterrestrial technology, I often wondered: What happened to our childhood curiosity and innocence? Throughout the media storm that descended on me as a result of my most public work (yet) on SETI, I was often animated by a simple thought: *If I attract one child somewhere in the world into science as a result of my answering the demands of the media, I will be satisfied. And if I make the public and, perhaps, even my profession more willing to entertain my unusual hypothesis, so much the better.*

• • •

In the spirit of the thought experiments with which I opened this book, the ones that I put to my Harvard undergraduates, here's another:

Imagine that back in, say, 1976, NASA uncovered proof of extraterrestrial life on another planet—say, Mars. NASA had sent a probe to the Red Planet; this probe conducted soil samples that were analyzed and found to contain evidence of life. And the result was that the ultimate question—is terrestrial life the only life in the universe?—had been clearly answered. The data was presented by the scientific community and embraced by the public.

As a result, for the past forty years, humanity has gone about its daily activities and scientific explorations with the understanding that there is nothing unique about terrestrial life, because if evidence of life exists on Mars, it is a near statistical certainty that it exists elsewhere. Guided by that understanding, the committees that evaluate and fund new scientific undertakings and instruments have decided to channel money toward the further search for life beyond Earth. Public funds have flowed to support these new explorations. Textbooks have been rewritten, graduate programs refocused, old presumptions challenged.

And now imagine that forty years after evidence of organic life was found on Mars, a small interstellar object—highly luminous, oddly tumbling, with a 91 percent probability of being disk-shaped

—passed through our solar system and, without visible outgassing, smoothly accelerated from a path that deviated from the force of the Sun's gravity alone, with an extra push that declined inversely with distance squared.

And now imagine that astronomers gleaned enough data about this object to understand these anomalies and that a few scientists studied the data and declared that one possible explanation for this object's peculiar features was that it was of extraterrestrial origin.

What, in this alternate reality, do you think would have been the profession's and the public's reaction to such a hypothesis?

With forty years to accommodate itself to evidence of extraterrestrial life, the world, I suspect, would have viewed this hypothesis as less exotic among all the unusual scenarios being offered to explain 'Oumuamua's peculiarities. Perhaps the world would also have spent the intervening forty years organizing itself in a way so that it was better prepared to detect and study 'Oumuamua; this could have allowed scientists to detect 'Oumuamua as early as July 2017, which would have given them sufficient time to launch a spacecraft to meet this peculiar object and photograph its surface from a close distance.

And perhaps, rather than awaiting the date when the Starshot Initiative sends its first lightsail craft out into the universe, as we are now, we would be awaiting the imminent return of the data from those very craft that were launched twenty years ago.

This thought experiment has two purposes. It should remind us that while we cannot control the data that the universe provides, we can control how we go about seeking it, assessing it, and recalibrating our future scientific undertakings. The world of possibilities to which we choose to open ourselves, bounded by the evidence we collect and that we allow our collective intelligence to consider, very much determines the world in which our children and grandchildren will live.

The second purpose of this thought experiment is to highlight an opportunity missed.

In 1975, NASA sent two Viking landers to Mars; these small probes arrived on the Red Planet the following year. They conducted experiments and collected soil samples, which they then analyzed. All of the results were transmitted back to Earth.

In October of 2019, Gilbert V. Levin—principal investigator on one of the experiments conducted by the Viking landers, the "labeled release" experiment—published an article in *Scientific American* stating the experiment had produced positive results that were proof of life on Mars. Designed to test Martian soil for evidence of life, Levin wrote in the article, "It seemed we had answered that ultimate question."

The experiment was simple: Introduce nutrients to the Martian soil and see if anything in that soil consumed it as food. The lander was equipped with radioactivity monitors capable of detecting any traces of metabolism such consumption would produce. What is more, the lander could repeat the experiment after heating the soil to a degree that would kill off all known life. If in the first experiment there was evidence of metabolism and in the second experiment there was none, that would suggest biological life at work.

And that, according to Levin, is precisely what the experiment showed.

Other experiments, however, failed to find corroborating evidence of life on Mars. The result of the first was deemed by NASA to have been a false positive. And in the decades since, no subsequent NASA lander to Mars has included instruments to follow up on that experiment.

NASA and other space agencies are planning to land rovers on Mars with equipment designed to search for past signs of life. Appropriately, the device on NASA's rover is named the Scanning Habitable Environments with Raman and Luminescence for Organics and Chemicals instrument—SHERLOC, for short. We can all take a measure of comfort from the fact that, however haltingly, the detective work of science continues.

Afterword

IN THE COURTROOM, CORROBORATING EYEWITness testimonies can make the difference between a defendant being acquitted or spending the rest of their life in jail. But in science, such testimonies are of decidedly more limited value.

This important distinction has historically held back the hunt for extraterrestrial intelligence — particularly in the context of Unidentified Flying Objects (UFOs), also referred to as Unidentified Aerial Phenomena (UAP). Past reports about UAP constitute a mixed bag, containing eyewitness testimonies and low-quality instrumental data. This evidence is next to useless, scientifically speaking, because science mandates quantitative measurements by instruments, removing the subjective impressions of humans from the equation. In addition, science rests on reproducible results which can be replicated by arranging similar circumstances.

In part because solid evidence about UAP has been in such short supply, the issue itself — and our ignorance about it — has grown more urgent. On June 25, 2021, the Office of the Director of

National Intelligence delivered a UAP report to the United States Congress admitting that some UAP are real objects of an unknown nature. Is this class of unidentified aerial phenomena of terrestrial origin, and is it a U.S. national security threat — for instance, some sort of advanced technology developed by known adversaries of the United States, such as Russia or China? Or are these objects instead the product of some other maker?

Rather than dismiss the UAP evidence as insufficient, scientists should replicate it with better instruments. This is the rationale behind the Galileo Project, a scientific endeavor that I initiated a month after the Pentagon report.

The raison d'être of the Galileo Project is to scientifically explore the nature of UAP and ʻOumuamua-like objects, and to gather high-quality data that will hopefully eliminate doubts about — and help us to discern the true nature of — these objects. A megapixel image of one of them will be worth thousands of words, so that is what we are aiming to capture: the data from a system of optical, infrared, or radio telescopes will be fed to state-of-the-art cameras, linked to a computer system with artificial intelligence software that will filter out objects of interest for the telescopes to track. Natural phenomena, such as birds or lightning, which are of interest to zoologists, or human-made equipment, such as drones or airplanes, which are of interest to some residents of Washington, DC, will not be studied by the Galileo Project. Rather, it will be a fishing expedition for a rarer type of target: objects whose appearances do not have mundane explanations.

Confronted with a scientific mystery, we chose to look through telescopes for answers. What we will find, only time will tell. But even if the only result of our efforts is to encourage a new generation of scientists to accept Galileo's challenge, his eponymous project will have been a success.

September 2, 2021

Acknowledgments

My deepest gratitude goes to my parents, Sara and David, who wisely encouraged my curiosity and wonder throughout my never-ending childhood, and to my remarkable wife, Ofrit, and our stunning daughters, Klil and Lotem, whose unconditional support and love make my life worth living.

Throughout my scientific career, I benefited greatly from collaborations with dozens of brilliant students and postdocs, a small fraction of whom are mentioned by name throughout the book and whose full scope of work can be found on my website, https://www.cfa.harvard.edu/~loeb/. As Rabbi Hanina remarked in the Talmud: "I have learned much from my teachers, more from my colleagues, and the most from my students."

This book would never have been written without key team members. In particular, I am grateful to my literary agents, Leslie Meredith and Mary Evans, for convincing me to write this book in the midst of a hectic research schedule; to editors Alex Littlefield and Georgina Laycock for their generous support and advice on this writing project; and to Thomas LeBien and Amanda Moon for their extraordinarily professional and brilliant insights in assembling and organizing the materials of this book. I also thank Michael Lemonick, the editor of the *Scientific American* blog *Observations*, for providing me with an invaluable platform for developing my opinions and arguments.

This assembly of helpful collaborators taught me what I know about myself and hence about the world. After all, the horizons of the universe we discover are set by what we imagine exists out there.

Notes

1. SCOUT

page

3 *"a messenger from afar"*: International Astronomical Union, "The IAU Approves New Type of Designation for Interstellar Objects," November 14, 2017, https://www.iau.org/news/announcements/detail/ann17045/.

3. ANOMALIES

33 *its trajectory deviated from what was expected:* Marco Micheli et al., "Non-Gravitational Acceleration in the Trajectory of 1I/2017 U1 ('Oumuamua)," *Nature* 559 (2018): 223–26, https://www.ifa.hawaii.edu/~meech/papers/2018/Micheli2018-Nature.pdf.

37 *changed the tumbling period of `Oumuamua:* Roman Rafikov, "Spin Evolution and Cometary Interpretation of the Interstellar Minor Object 1I/2017 `Oumuamua," *Astrophysical Journal* (2018), https://arxiv.org/pdf/1809.06389.pdf.

38 *"did not detect the object"*: David E. Trilling et al., "Spitzer Observations of Interstellar Object 1I/'Oumuamua," *Astronomical Journal* (2018), https://arxiv.org/pdf/1811.08072.pdf.

39 *astronomers reviewed images:* Man-To Hui and Mathew M. Knight, "New In-

sights into Interstellar Object 1I/2017 U1 ('Oumuamua) from SOHO/STEREO Nondetections," *Astronomical Journal* (2019), https://arxiv.org/pdf/1910.10303.pdf.

"the greatest comet-finder": NASA, "Nearing 3,000 Comets: SOHO Solar Observatory Greatest Comet Hunter of All Time," July 30, 2015, https://www.nasa.gov/feature/goddard/soho/solar-observatory-greatest-comet-hunter-of-all-time.

'Oumuamua's ice was entirely made of hydrogen: Darryl Seligman and Gregory Laughlin, "Evidence That 1I/2017 U1 ('Oumuamua) Was Composed of Molecular Hydrogen Ice," *Astrophysical Journal Letters* (2020), https://arxiv.org/pdf/2005.12932.pdf.

43 *"a devolatilized aggregate":* Zdenek Sekanina, "1I/'Oumuamua As Debris of Dwarf Interstellar Comet That Disintegrated Before Perihelion," arXiv.org (2019), https://arxiv.org/pdf/1901.08704.pdf.

A similar concept: Amaya Moro-Martin, "Could 1I'Oumuamua Be an Icy Fractal Aggregate," *Astrophysical Journal* (2019), https://arxiv.org/pdf/1902.04100.pdf.

44 *went back to the evidence:* Sergey Mashchenko, "Modeling the Light Curve of 'Oumuamua: Evidence for Torque and Disk-Like Shape," *Monthly Notices of the Royal Astronomical Society* (2019), https://arxiv.org/pdf/1906.03696.pdf.

45 *melting and tidal stretching along the trajectory:* Yun Zhang and Douglas N. C. Lin, "Tidal Fragmentation as the Origin of 1I/2017 U1 ('Oumuamua)," *Nature Astronomy* (2020), https://arxiv.org/pdf/2004.07218.pdf.

5. THE LIGHTSAIL HYPOTHESIS

69 *"we find no compelling evidence":* 'Oumuamua ISSI Team, "The Natural History of 'Oumuamua," *Nature Astronomy* 3 (2019), https://arxiv.org/pdf/1907.01910.pdf.

"We have never seen anything like 'Oumuamua": Michelle Starr, "Astronomers Have Analysed Claims 'Oumuamua's an Alien Ship, and It's Not Looking Good," *Science Alert,* July 1, 2019, https://www.sciencealert.com/astronomers-have-determined-oumuamua-is-really-truly-not-an-alien-lightsail.

6. SEASHELLS AND BUOYS

76 *they reached some general conclusions:* Aaron Do, Michael A. Tucker, and John Tonry, "Interstellar Interlopers: Number Density and Origin of 'Oumuamua-Like Objects," *Astrophysical Journal* (2018), https://arxiv.org/pdf/1801.02821.pdf.

77 *In two follow-up papers:* Amaya Moro-Martin, "Origin of 1I'Oumuamua. I. An Ejected Protoplanetary Disk Object?," *Astrophysical Journal* (2018),

https://arxiv.org/pdf/1810.02148.pdf; Amaya Moro-Martin, "II. An Ejected Exo-Oort Cloud Object," *Astronomical Journal* (2018), https://arxiv.org/pdf/1811.00023.pdf.

81 *only one in every five hundred stars:* Eric Mamajek, "Kinematics of the Interstellar Vagabond 1I/'Oumuamua (A/2017 U1)," *Research Notes of the American Astronomical Society* (2017), https://arxiv.org/abs/1710.11364.

7. LEARNING FROM CHILDREN

90 *two simple conjectures:* Giuseppe Cocconi and Philip Morrison, "Searching for Interstellar Communications," *Nature* 184, no. 4690 (September 19, 1959): 844–46, http://www.iaragroup.org/_OLD/seti/pdf_IARA/cocconi.pdf.

93 *"Millions have been spent":* Adam Mann, "Intelligent Ways to Search for Extraterrestrials," *New Yorker* (October 3, 2019).

107 *currently on track:* Jason Wright, "SETI Is a Very Young Field (Academically)," *AstroWright* (blog), January 23, 2019, https://sites.psu.edu/astrowright/2019/01/23/seti-is-a-very-young-field-academically/.

9. FILTERS

122 *the World Bank issued a report:* Silpa Kaza et al., "What a Waste 2.0: A Global Snapshot of Solid Waste Management to 2050," World Bank (2018), https://openknowledge.worldbank.org/handle/10986/30317.

128 *"I am not sufficiently conceited":* Mario Livio, "Winston Churchill's Essay on Alien Life Found," *Nature* (2017), https://www.nature.com/news/winston-churchill-s-essay-on-alien-life-found-1.21467; Brian Handwerk, "'Are We Alone in the Universe?' Winston Churchill's Lost Extraterrestrial Essay Says No," SmithsonianMag.com, February 16, 2017, https://www.smithsonianmag.com/science-nature/winston-churchill-question-alien-life-180962198/.

13. SINGULARITIES

184 *"Last week in New York":* You can watch a video of Hawking's brief speech at my family's home on April 22, 2016, at this link: https://www.cfa.harvard.edu/~loeb/SI.html.

CONCLUSION

195 *"It seemed we had answered":* Gilbert V. Levin, "I'm Convinced We Found Evidence of Life on Mars in the 1970s," *Scientific American,* October 10, 2019, https://blogs.scientificamerican.com/observations/im-convinced-we-found-evidence-of-life-on-mars-in-the-1970s/.

Additional Reading

Many of the ideas covered in this book were first broached and explored in my previously published articles. A list, with hyperlinks, is available here: https://www.cfa.harvard.edu/~loeb/Oumuamua.html.

Below are some of my articles for additional enrichment on the topics of each chapter. All URLs provided for academic journal articles in this section and the next direct to the arXiv, a preprint server that makes academic papers available to the scientific community and the general public.

INTRODUCTION

Loeb, A. "The Case for Cosmic Modesty." *Scientific American*, June 28, 2017, https://blogs.scientificamerican.com/observations/the-case-for-cosmic-modesty/.

———. "Science Is Not About Getting More Likes." *Scientific American*, October 8, 2019, https://blogs.scientificamerican.com/observations/science-is-not-about-getting-more-likes/.

———. "Seeking the Truth When the Consensus Is Against You." *Scientific American*,

November 9, 2018, https://blogs.scientificamerican.com/observations/seeking-the-truth-when-the-consensus-is-against-you/.
——. "Essential Advice for Fledgling Scientists." *Scientific American*, December 2, 2019, https://blogs.scientificamerican.com/observations/essential-advice-for-fledgling-scientists/.
——. "A Tale of Three Nobels." *Scientific American*, December 18, 2019, https://blogs.scientificamerican.com/observations/a-tale-of-three-nobels/.
——. "Advice to Young Scientists: Be a Generalist." *Scientific American*, March 16, 2020, https://blogs.scientificamerican.com/observations/advice-for-young-scientists-be-a-generalist/.
——. "The Power of Scientific Brainstorming." *Scientific American*, July 23, 2020, https://www.scientificamerican.com/article/the-power-of-scientific-brainstorming/.
——. "A Movie of the Evolving Universe Is Potentially Scary." *Scientific American*, August 2, 2020. https://www.scientificamerican.com/article/a-movie-of-the-evolving-universe-is-potentially-scary/.
Moro-Martin, A., E. L. Turner, and A. Loeb. "Will the Large Synoptic Survey Telescope Detect Extra-Solar Planetesimals Entering the Solar System?" *Astrophysical Journal* (2009), https://arxiv.org/pdf/0908.3948.pdf.

1. SCOUT

Bialy, S., and A. Loeb. "Could Solar Radiation Pressure Explain 'Oumuamua's Peculiar Acceleration?" *Astrophysical Journal Letters* (2018), https://arxiv.org/pdf/1810.11490.pdf.
Loeb, A. "Searching for Relics of Dead Civilizations." *Scientific American*, September 27, 2018, https://blogs.scientificamerican.com/observations/how-to-search-for-dead-cosmic-civilizations/.
——. "Are Alien Civilizations Technologically Advanced?" *Scientific American*, January 8, 2018, https://blogs.scientificamerican.com/observations/are-alien-civilizations-technologically-advanced/.
——. "Q&A with a Journalist." Center for Astrophysics, Harvard University, January 25, 2019, https://www.cfa.harvard.edu/~loeb/QA.pdf.

2. THE FARM

Loeb, A. "The Humanities and the Future." *Scientific American*, March 22, 2019, https://blogs.scientificamerican.com/observations/the-humanities-and-the-future/.
——. "What Is the One Thing You Would Change About the World?" *Harvard Gazette*, July 1, 2019, https://news.harvard.edu/gazette/story/2019/06/focal-point-harvard-professor-avi-loeb-wants-more-scientists-to-think-like-children/.
——. "Science as a Way of Life." *Scientific American*, August 14, 2019, https://blogs

.scientificamerican.com/observations/a-scientist-must-go-where-the-evidence -leads/.

——. "Beware of Theories of Everything." *Scientific American,* June 9, 2020, https://blogs.scientificamerican.com/observations/beware-of-theories-of-everything/.

Loeb, A., and E. L. Turner. "Detection Technique for Artificially Illuminated Objects in the Outer Solar System and Beyond." *Astrobiology* (2012), https://arxiv.org/pdf/1110.6181.pdf.

3. ANOMALIES

Hoang, T., and A. Loeb. "Destruction of Molecular Hydrogen Ice and Implications for 1I/2017 U1 ('Oumuamua)." *Astrophysical Journal Letters* (2020), https://arxiv.org/pdf/2006.08088.pdf.

Lingam, M., and A. Loeb. "Implications of Captured Interstellar Objects for Panspermia and Extraterrestrial Life." *Astrophysical Journal* (2018), https://arxiv.org/pdf/1801.10254.pdf.

Loeb, A. "Theoretical Physics Is Pointless Without Experimental Tests." *Scientific American,* August 10, 2018, https://blogs.scientificamerican.com/observations/theoretical-physics-is-pointless-without-experimental-tests/.

——. "The Power of Anomalies." *Scientific American,* August 28, 2018, https://blogs.scientificamerican.com/observations/the-power-of-anomalies/.

——. "On 'Oumuamua." Center for Astrophysics, Harvard University, November 5, 2018, https://www.cfa.harvard.edu/~loeb/Oumuamua.pdf.

——. "Six Strange Facts About the First Interstellar Visitor, 'Oumuamua." *Scientific American,* November 20, 2018, https://blogs.scientificamerican.com/observations/6-strange-facts-about-the-interstellar-visitor-oumuamua/.

——. "How to Approach the Problem of 'Oumuamua." *Scientific American,* December 19, 2018, https://blogs.scientificamerican.com/observations/how-to-approach-the-problem-of-oumuamua/.

——. "The Moon as a Fishing Net for Extraterrestrial Life." *Scientific American,* September 25, 2019, https://blogs.scientificamerican.com/observations/the-moon-as-a-fishing-net-for-extraterrestrial-life/.

——. "The Simple Truth About Physics." *Scientific American,* January 1, 2020, https://blogs.scientificamerican.com/observations/the-simple-truth-about-physics/.

Sheerin, T. F., and A. Loeb. "Could 1I/2017 U1 'Oumuamua Be a Solar Sail Hybrid?" Center for Astrophysics, Harvard University, May 2020, https://www.cfa.harvard.edu/~loeb/TL.pdf.

Siraj, A., and A. Loeb. "'Oumuamua's Geometry Could Be More Extreme than Previously Inferred." *Research Notes of the American Astronomical Society* (2019), http://iopscience.iop.org/article/10.3847/2515-5172/aafe7c/meta.

——. "Identifying Interstellar Objects Trapped in the Solar System Through Their Orbital Parameters." *Astrophysical Journal Letters* (2019), https://arxiv.org/pdf/1811.09632.pdf.

——. "An Argument for a Kilometer-Scale Nucleus of C/2019 Q4." *Research Notes of the American Astronomical Society* (2019), https://arxiv.org/pdf/1909.07286.pdf.

4. STARCHIPS

Christian, P., and A. Loeb. "Interferometric Measurement of Acceleration at Relativistic Speeds." *Astrophysical Journal* (2017), https://arxiv.org/pdf/1608.08230.pdf.

Guillochon, J., and A. Loeb. "SETI via Leakage from Light Sails in Exoplanetary Systems." *Astrophysical Journal* (2016), https://arxiv.org/pdf/1508.03043.pdf.

Kreidberg, L., and A. Loeb. "Prospects for Characterizing the Atmosphere of Proxima Centauri b." *Astrophysical Journal Letters* (2016), https://arxiv.org/pdf/1608.07345.pdf.

Loeb, A. "On the Habitability of the Universe." *Consolidation of Fine Tuning* (2016), https://arxiv.org/pdf/1606.08926.pdf.

——. "Searching for Life Among the Stars." *Pan European Networks: Science and Technology*, July 2017, https://www.cfa.harvard.edu/~loeb/PEN.pdf.

——. "Breakthrough Starshot: Reaching for the Stars." *SciTech Europa Quarterly*, March 2018, https://www.cfa.harvard.edu/~loeb/Loeb_Starshot.pdf.

——. "Sailing on Light." *Forbes*, August 8, 2018, https://www.cfa.harvard.edu/~loeb/Loeb_Forbes.pdf.

——. "Interstellar Escape from Proxima b Is Barely Possible with Chemical Rockets." *Scientific American*, 2018, https://arxiv.org/pdf/1804.03698.pdf.

Loeb, A., R. A. Batista, and D. Sloan. "Relative Likelihood for Life as a Function of Cosmic Time." *Journal of Cosmology and Astroparticle Physics* (2016), https://arxiv.org/pdf/1606.08448.pdf.

Manchester, Z., and A. Loeb. "Stability of a Light Sail Riding on a Laser Beam." *Astrophysical Journal Letters* (2017), https://arxiv.org/pdf/1609.09506.pdf.

5. THE LIGHTSAIL HYPOTHESIS

Hoang, T., and A. Loeb. "Electromagnetic Forces on a Relativistic Spacecraft in the Interstellar Medium." *Astrophysical Journal* (2017), https://arxiv.org/pdf/1706.07798.pdf.

Hoang, T., A. Lazarian, B. Burkhart, and A. Loeb. "The Interaction of Relativistic Spacecrafts with the Interstellar Medium." *Astrophysical Journal* (2017), https://arxiv.org/pdf/1608.05284.pdf.

Hoang, T., A. Loeb, A. Lazarian, and J. Cho. "Spinup and Disruption of Interstellar Asteroids by Mechanical Torques, and Implications for 1I/2017 U1 ('Oumuamua)." *Astrophysical Journal* (2018), https://arxiv.org/pdf/1802.01335.pdf.

Loeb, A. "An Audacious Explanation for Fast Radio Bursts." *Scientific American*, June 24, 2020, https://www.scientificamerican.com/article/an-audacious-explanation-for-fast-radio-bursts/.

6. SEASHELLS AND BUOYS

Lingam, M., and A. Loeb. "Risks for Life on Habitable Planets from Superflares of Their Host Stars." *Astrophysical Journal* (2017), https://arxiv.org/pdf/1708.04241.pdf.

——. "Optimal Target Stars in the Search for Life." *Astrophysical Journal Letters* (2018), https://arxiv.org/pdf/1803.07570.pdf.

Loeb, A. "For E.T. Civilizations, Location Could Be Everything." *Scientific American,* March 13, 2018, https://blogs.scientificamerican.com/observations/for-e-t-civilizations-location-could-be-everything/.

——. "Space Archaeology." *Atmos,* November 8, 2019, https://www.cfa.harvard.edu/~loeb/Atmos_Loeb.pdf

Siraj, A., and A. Loeb. "Radio Flares from Collisions of Neutron Stars with Interstellar Asteroids." *Research Notes of the American Astronomical Society* (2019), https://arxiv.org/pdf/1908.11440.pdf.

——. "Observational Signatures of Sub-Relativistic Meteors." *Astrophysical Journal Letters* (2020), https://arxiv.org/pdf/2002.01476.pdf.

7. LEARNING FROM CHILDREN

Lingam, M., and A. Loeb. "Fast Radio Bursts from Extragalactic Light Sails." *Astrophysical Journal Letters* (2017), https://arxiv.org/pdf/1701.01109.pdf.

——. "Relative Likelihood of Success in the Searches for Primitive Versus Intelligent Life." *AstroBiology* (2019), https://arxiv.org/pdf/1807.08879.pdf.

8. VASTNESS

Loeb, A. "Geometry of the Universe." *Astronomy,* July 8, 2020, https://www.cfa.harvard.edu/~loeb/Geo.pdf.

——. *How Did the First Stars and Galaxies Form?* Princeton, NJ: Princeton University Press, 2010.

Loeb, A., and S. R. Furlanetto. *The First Galaxies in the Universe.* Princeton, NJ: Princeton University Press, 2013.

Loeb, A., and M. Zaldarriaga. "Eavesdropping on Radio Broadcasts from Galactic Civilizations with Upcoming Observatories for Redshifted 21 Cm Radiation." *Journal of Cosmology and Astroparticle Physics* (2007), https://arxiv.org/pdf/astro-ph/0610377.pdf.

9. FILTERS

Lingam, M., and A. Loeb. "Propulsion of Spacecrafts to Relativistic Speeds Using Natural Astrophysical Sources." *Astrophysical Journal* (2020), https://arxiv.org/pdf/2002.03247.pdf.

Loeb, A. "Our Future in Space Will Echo Our Future on Earth." *Scientific American,* January 10, 2019, https://blogs.scientificamerican.com/observations/our-future-in-space-will-echo-our-future-on-earth/.

——. "When Lab Experiments Carry Theological Implications." *Scientific American,* April 22, 2019, https://blogs.scientificamerican.com/observations/when-lab-experiments-carry-theological-implications/.

——. "The Only Thing That Remains Constant Is Change." *Scientific American,* September 6, 2019, https://blogs.scientificamerican.com/observations/the-only-thing-that-remains-constant-is-change/.

Siraj, A., and A. Loeb. "Exporting Terrestrial Life Out of the Solar System with Gravitational Slingshots of Earthgrazing Bodies." *International Journal of Astrobiology* (2019), https://arxiv.org/pdf/1910.06414.pdf.

10. ASTRO-ARCHAEOLOGY

Lin, H. W., G. Gonzalez Abad, and A. Loeb. "Detecting Industrial Pollution in the Atmospheres of Earth-Like Exoplanets." *Astrophysical Journal Letters* (2014), https://arxiv.org/pdf/1406.3025.pdf.

Lingam, M., and A. Loeb. "Natural and Artificial Spectral Edges in Exoplanets." *Monthly Notices of the Royal Astronomical Society* (2017), https://arxiv.org/pdf/1702.05500.pdf.

Loeb, A. "Making the Church Taller." *Scientific American,* October 18, 2018, https://blogs.scientificamerican.com/observations/making-the-church-taller/.

——. "Advanced Extraterrestrials as an Approximation to God." *Scientific American,* January 26, 2019, https://blogs.scientificamerican.com/observations/advanced-extraterrestrials-as-an-approximation-to-god/.

——. "Are We Really the Smartest Kid on the Cosmic Block?" *Scientific American,* March 4, 2019, https://blogs.scientificamerican.com/observations/are-we-really-the-smartest-kid-on-the-cosmic-block/.

——. "Visionary Science Takes More Than Just Technical Skills." *Scientific American,* May 25, 2020, https://blogs.scientificamerican.com/observations/visionary-science-takes-more-than-just-technical-skills/.

11. 'OUMUAMUA'S WAGER

Chen, H., J. C. Forbes, and A. Loeb. "Habitable Evaporated Cores and the Occurrence of Panspermia near the Galactic Center." *Astrophysical Journal Letters* (2018), https://arxiv.org/pdf/1711.06692.pdf.

Cox, T. J., and A. Loeb. "The Collision Between the Milky Way and Andromeda." *Monthly Notices of the Royal Astronomical Society* (2008), https://arxiv.org/pdf/0705.1170.pdf.

Forbes, J. C., and A. Loeb. "Evaporation of Planetary Atmospheres Due to XUV Illumination by Quasars." *Monthly Notices of the Royal Astronomical Society* (2018), https://arxiv.org/pdf/1705.06741.pdf.

Loeb, A. "Long-Term Future of Extragalactic Astronomy." *Physical Review D* (2002), https://arxiv.org/pdf/astro-ph/0107568.pdf.

——. "Cosmology with Hypervelocity Stars." *Journal of Cosmology and Astroparticle Physics* (2011), https://arxiv.org/pdf/1102.0007.pdf.

——. "Why a Mission to a Visiting Interstellar Object Could Be Our Best Bet for Finding Aliens." *Gizmodo,* October 31, 2018, https://gizmodo.com/why-a-mission-to-a-visiting-interstellar-object-could-b-1829975366.

——. "Be Kind to Extraterrestrials." *Scientific American,* February 15, 2019, https://blogs.scientificamerican.com/observations/be-kind-to-extraterrestrials/.

——. "Living Near a Supermassive Black Hole." *Scientific American,* March 11, 2019, https://blogs.scientificamerican.com/observations/living-near-a-supermassive-black-hole/.

12. SEEDS

Ginsburg, I., M. Lingam, and A. Loeb. "Galactic Panspermia." *Astrophysical Journal* (2018), https://arxiv.org/pdf/1810.04307.pdf.

Lingam, M., I. Ginsburg, and A. Loeb. "Prospects for Life on Temperate Planets Around Brown Dwarfs." *Astrophysical Journal* (2020), https://arxiv.org/pdf/1909.08791.pdf.

Lingam, M., and A. Loeb. "Subsurface Exolife." *International Journal of Astrobiology* (2017), https://arxiv.org/pdf/1711.09908.pdf.

——. "Brown Dwarf Atmospheres as the Potentially Most Detectable and Abundant Sites for Life." *Astrophysical Journal* (2019), https://arxiv.org/pdf/1905.11410.pdf.

——. "Dependence of Biological Activity on the Surface Water Fraction of Planets." *Astronomical Journal* (2019), https://arxiv.org/pdf/1809.09118.pdf.

——. "Physical Constraints for the Evolution of Life on Exoplanets." *Reviews of Modern Physics* (2019), https://arxiv.org/pdf/1810.02007.pdf.

Loeb, A. "In Search of Green Dwarfs." *Scientific American,* June 3, 2019, https://blogs.scientificamerican.com/observations/in-search-of-green-dwarfs/.

——. "Did Life from Earth Escape the Solar System Eons Ago?" *Scientific American,* November 4, 2019, https://blogs.scientificamerican.com/observations/did-life-from-earth-escape-the-solar-system-eons-ago/.

——. "What Will We Do When the Sun Gets Too Hot for Earth's Survival?" *Scientific American,* November 25, 2019, https://blogs.scientificamerican.com/observations/what-will-we-do-when-the-sun-gets-too-hot-for-earths-survival/.

——. "Surfing a Supernova." *Scientific American,* February 3, 2020, https://blogs.scientificamerican.com/observations/surfing-a-supernova/.

Siraj, A., and A. Loeb. "Transfer of Life by Earth-Grazing Objects to Exoplanetary Systems." *Astrophysical Journal Letters* (2020), https://arxiv.org/pdf/2001.02235.pdf.

Sloan, D., R. A. Batista, and A. Loeb. "The Resilience of Life to Astrophysical Events." *Nature Scientific Reports* (2017), https://arxiv.org/pdf/1707.04253.pdf.

13. SINGULARITIES

Broderick, A., and A. Loeb. "Portrait of a Black Hole." *Scientific American*, 2009, https://www.cfa.harvard.edu/~loeb/sciam2.pdf.

Forbes, J., and A. Loeb. "Turning Up the Heat on 'Oumuamua." *Astrophysical Journal Letters* (2019), https://arxiv.org/pdf/1901.00508.pdf.

Loeb, A. "'Oumuamua's Cousin?" *Scientific American*, May 6, 2019, https://blogs.scientificamerican.com/observations/oumuamuas-cousin/.

——. "It Takes a Village to Declassify an Error Bar." *Scientific American*, July 3, 2019, https://blogs.scientificamerican.com/observations/it-takes-a-village-to-declassify-an-error-bar/.

——. "Can the Universe Provide Us with the Meaning of Life?" *Scientific American*, November 18, 2019, https://blogs.scientificamerican.com/observations/can-the-universe-provide-us-with-the-meaning-of-life/.

——. "In Search of Naked Singularities." *Scientific American*, May 3, 2020, https://blogs.scientificamerican.com/observations/in-search-of-naked-singularities/.

Siraj, A., and A. Loeb. "Discovery of a Meteor of Interstellar Origin." *Astrophysical Journal Letters* (2019), https://arxiv.org/pdf/1904.07224.pdf.

——. "Probing Extrasolar Planetary Systems with Interstellar Meteors." *Astrophysical Journal Letters* (2019), https://arxiv.org/pdf/1906.03270.pdf.

——. "Halo Meters." *Astrophysical Journal Letters* (2019), https://arxiv.org/pdf/1906.05291.pdf.

CONCLUSION

Lingam, M., and A. Loeb. "Searching the Moon for Extrasolar Material and the Building Blocks of Extraterrestrial Life." *Publications of the National Academy of Sciences* (2019), https://arxiv.org/pdf/1907.05427.pdf.

Loeb, A. "Science Is an Infinite-Sum Game." *Scientific American*, July 31, 2018, https://blogs.scientificamerican.com/observations/science-is-an-infinite-sum-game/.

——. "Why Should Scientists Mentor Students?" *Scientific American*, February 25, 2020, https://blogs.scientificamerican.com/observations/why-should-scientists-mentor-students/.

——. "Why the Pursuit of Scientific Knowledge Will Never End." *Scientific American*, April 6, 2020, https://blogs.scientificamerican.com/observations/why-the-pursuit-of-scientific-knowledge-will-never-end/.

——. "A Sobering Astronomical Reminder from COVID-19." *Scientific American*, April 22, 2020, https://blogs.scientificamerican.com/observations/a-sobering-astronomical-reminder-from-covid-19/.

——. "Living with Scientific Uncertainty." *Scientific American*, July 15, 2020, https://www.scientificamerican.com/article/living-with-scientific-uncertainty/.

——. "What If We Could Live for a Million Years?" *Scientific American*, August 16, 2020, https://www.scientificamerican.com/article/what-if-we-could-live-for-a-million-years/.

Siraj, A., and A. Loeb. "Detecting Interstellar Objects through Stellar Occultations." *Astrophysical Journal* (2019), https://arxiv.org/pdf/2001.02681.pdf.

——. "A Real-Time Search for Interstellar Impact on the Moon." *Acta Astronautica* (2019), https://arxiv.org/pdf/1908.08543.pdf.

Index